THE WIND POWER BOOK

The Wind Power Book

by **Jack Park**

with a foreword by
Robert Redford

Cheshire Books, Palo Alto, California

Copyright ©1981 by Jack Park
Illustrations copyright ©1981 by Cheshire Books

Printed in the United States of America

Published by Cheshire Books
514 Bryant Street, Palo Alto, CA 94301, U.S.A.

Distributed in the U.S.A. by Van Nostrand Reinhold
135 West 50th Street, New York, NY 10020, U.S.A.

Distributed in Canada by Firefly Books, Ltd.
3520 Pharmacy Avenue, Scarborough,
Ontario M1W 2T8

Distributed in Great Britain by Prism Press
Stable Court, Chalmington, Dorchester,
Dorset DT2 0HB

Distributed in Australia by Second Back Row Press
50 Govett Street, Katoomba, N.S.W. 2780

12 11 10 9 8 7 6 5 4 3 2

Credits:

Michael Riordan: Editor
Linda Goodman: Art direction, book design
Edward Wong-Ligda: Illustrations, cover and
    book design
Sara Boore: Graphs, charts and maps
R. G. Beukers: Copy editing, proofreading
Deborah Wong: Production assistant
Robert Cooney: Paste-up

Cover Photo (by Mick Csaky and Heini Schneebeli):
A Darrieus rotor wind turbine at the Centre for
Alternative Technology, Machynlleth, Wales,
United Kingdom.

Library of Congress Cataloging in Publication Data

Park, Jack.
    The wind power book.

    Bibliography: p.
    Includes index.
    1. Wind power.   I. Title.
TK1541.P37   621.4'5   81-128
ISBN 0-917352-05-X   AACR2
ISBN 0-917352-06-8 (pbk.)

Publisher's Acknowledgments

The production of an illustrated book from the collection of words, photographs and sketches called a manuscript is a long journey, one that entails important contributions by many people. Often their hard work can go unrecognized— except for the gratitude of author and publisher. Cheshire Books wishes to thank the following individuals and firms here for their effort, patience and dedication in producing *The Wind Power Book:* Rain Blockley, final proofing; Susan Riggs, index; Patti Melt, back cover design; Harry Guiremand and Russell Mills, manuscript type; Lauren Langford, tables type; Robin Woodward, graphs; General Graphic Service, photography; Peninsula Blueprint Company, photostats; Frank's Type, display type. Our special thanks go to Edward Wong-Ligda, Deborah Wong, Sara Boore, R.G. Beukers, and Robert Cooney whose talent and persistence helped make this book a reality.

The following individuals and institutions provided photographs for this book. The page numbers and, where necessary, position (*t*-top, *b*-bottom, *l*-left, and *r*-right) of each photograph are listed before the source. All other photos are by Helen and Jack Park.

Cover inset: Mick Csaky and Heini Schneebeli; 15: Stanton D. Dornbirer; 36, 37*t*, 37*b*: Jim Cullen; 40, 41: Marshall Merriam; 44: NOAA/NESS; 61, 62*b*, 63: Helion, Inc.; 76: Sandia Laboratories; 84, back cover: Automatic Power Division of the Pennwalt Corporation; 115: Joe Carter; 116*l*, 116*r*: Enertech Corp.; 123: California Historical Society.

AUTHOR'S NOTE

My experience in wind power began in the late 1960's with the design of a small wind generator for powering my ranch in northern California. I soon discovered that little, if any, useful literature was available to assist inexperienced wind-machine designers. With the exception of books by Golding and Putnam, there were no readily available texts in the local libraries—and few in college libraries or special collections.

In the early 1970's, I organized my research notes into a small book called *Simplified Wind Power Systems for Experimenters.* Published by Helion, Inc., the book has gone through a second edition and been distributed to almost every country in the world. From its readers, I have received thousands of letters—each with a request for further information or a suggested change or improvement. Based upon this experience, I began to reorganize my notes and incorporated many of the lessons learned from earlier versions. *The Wind Power Book* is the end result of all this effort.

The wind power field has witnessed tremendous growth in the decade since I began to write *Simplified.* I have learned many new things—virtually all of them through the efforts of colleagues and friends. The most influential colleagues have been Dr.'s Peter Lissaman, Richard Schwind, and Ulrich Hütter. My most helpful friends and associates have been Bill Goddard, W.C. Strumpell, Ken Johnson, and (the late) Richard Dehr. All that I have learned in these years has been distilled into *The Wind Power Book.*

The book would not be a reality without the capable editorial assistance of Michael Riordan, the artistic talents of Edward Wong-Ligda, and the graphic design work of Linda Goodman. With all of this effort, perhaps the most credit should go to people who made the endless, tedious contributions to the production effort. Certainly the most important of these contributions has come from Helen Ann Park, my wife, who typed the three versions of the manuscript necessary to produce this book.

In writing this book, I have tried to reduce complex mathematics into simple graphs and arithmetic problems that will allow innovative people to use the fundamentals an engineer has available. Simple examples illustrate each step in the wind-machine design process. The approach used is not one of exact science, but rather one of approximation—using the best possible guesses and estimates. The numbers may not be exact, but they are usually well within the necessary accuracy. Some people will probably need a few machines "under their belts" before their calculations become sufficiently accurate.

As often happens, some places on earth are not well blessed with wind energy. Brownsville is one such place. My annual average windspeed falls below 8 mph—with an almost unmeasurable mean power density. Virtually no watts-per-square-meter available. But despite the lack of a useful wind resource, I installed a variety of wind machines at my ranch. Each summer evening, a fresh, useful breeze rises to help the water-pumpers keep my tanks full. The various windchargers I have installed, dismantled, reinstalled and redismantled have mostly stood as idle monuments to the wind I wished were available. I heartily recommend the field of wind power to you with one note of caution: all the engineering know-how and the wisdom gleaned from experience just cannot make the wind blow.

Jack Park
Brownsville, California
January, 1981

TABLE OF CONTENTS

FOREWORD

The Wind Power Book makes wind power accessible to the average person, but not for the reason that readers might expect. It avoids the uncritical approach to alternative energy that the public has seen so often and grown to mistrust. Instead, the book explains the technology as well as the theory of wind power in a realistic way.

Jack Park admits that wind power may not be the best alternative for everyone, and I believe this approach will inspire confidence in his readers. His trial-and-error style of reporting, the analysis of potential problems, and the discussions of cost-effectiveness, among other things, make it clear that Park knows what he's talking about. He doesn't try to oversell his subject. He also mentions the likelihood that wind power users will have to budget their energy use to match the system, and may even need a mix of energy sources—including the local utility company. Park is not an unrealistic purist; he knows he is not offering the be-all and end-all to our energy problem. What he does offer is a practical approach to wind power as just one of many possible alternative energy solutions.

Once readers discover that they are not expected to be zealots, they begin to recognize wind power as a potential solution to their own energy problems. Park shows them how they can determine its viability for their own needs. He gives practical guidance for surveying the winds at a site to see whether it's feasible, lists the characteristics of various wind machines to help people make informed choices, and gives common-sense hints on avoiding the many possible pitfalls.

One hesitates to mention the term "energy crisis" these days. But the term and the problems to which it alludes are still with us, and answers don't seem to be forthcoming from the usual institutional sources on which we have all come to rely. Books like *The Wind Power Book* let us know that there are other sources to which we can turn: the voices of informed writers like Jack Park and the voice of self reliance within ourselves.

Robert Redford
Provo, Utah
December, 1980

1

Introduction

**Harnessing the Wind
for Power and Energy**

So you want to design and build a wind power system? Maybe you're tired of paying ever-increasing electric bills and worried about the future availability of electricity as fossil fuels become depleted. Maybe you've just bought land far away from the nearest power line and you'd like to harness the wind to pump water for your cattle. Or maybe you're a New Age entrepreneur who plans to generate electrical power at several windy sites and sell it to the utilities. If so, you are entering the ranks of a growing number of people turning back to one of the oldest sources of energy and power.

The Egyptians are believed to be the first to make practical use of wind power. Around 2800 BC, they began to use sails to assist the rowing power of slaves. Eventually, sails assisted their draft animals in such tasks as grinding grain and lifting water.

The Persians began using wind power a few centuries before Christ, and by 700 AD they were building vertical-shaft windmills, or *panemones,* to power their grain-grinding stones. Other Mideast civilizations, most notably the Moslems, picked up where the Persians left off and built their own panemones. Returning Crusaders are thought to have brought windmill ideas and designs to Europe, but it was probably the Dutch who developed the horizontal-shaft, propellor-type windmills common to the Dutch and English countrysides. Wind and water power soon became the prime sources of mechanical energy in medieval England. During this period, the Dutch relied on wind power for

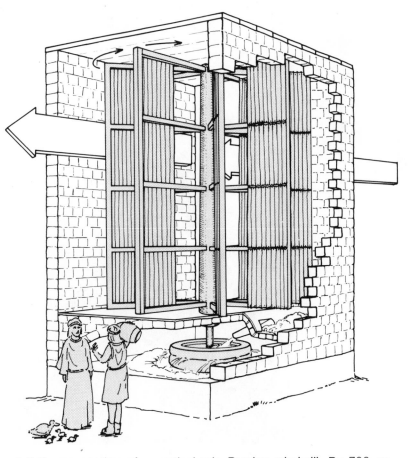

Artist's conception of a vertical-axis Persian windmill. By 700 AD, machines like this were grinding grain in the Middle East.

water pumping, grain grinding and sawmill operation.

Throughout the Middle Ages, technical improvements continued to occur in such areas as blade aerodynamics, gear design and the overall design of the windmill. The oldest European machines were the "post"

The multibladed windmill invented by Daniel Halliday. Developed in the mid-nineteenth century, this machine was the forerunner of the American Farm Windmill still seen in many rural areas today.

Grain-grinding windmill of Colonial New England. These early American mills were adapted from English and Dutch designs.

windmills—with the propellor assembly mounted atop a post set in the ground. The post served as a pivot that allowed the millwright to aim his windmill into the prevailing wind. Post windmill designs soon evolved into the "cap" windmill, in which the aiming

bearing was part of a cap that held the blades. People had to aim their wind machines until the invention, in 1750, of the fantail—an automatic aiming device powered by the wind itself.

The earliest windmills in Colonial New England were duplicates of English machines. Many of the design improvements used in Holland were virtually ignored. By 1850, Daniel Halliday began developing what was to become the famous American Farm Windmill. Used principally for pumping water, this machine is the familiar multibladed windmill still seen in many rural areas. Even today, cattle ranching would not be possible in many parts of America, Argentina and Australia without this machine.

Wind-generated electricity began around the turn of the century, with some of the earliest developments credited to the Danes. By the 1930's, about a dozen American firms were making and selling these *windchargers,* mostly to farmers in the windy Great Plains. Typically, these machines could deliver up to 1,000 watts (1 kilowatt, or 1kW) of direct current (DC) electrical power when the wind was blowing. Then came the Rural Electrification Administration (REA), a government-subsidized program to extend power lines to remote farms and homesteads. Windchargers were no match for the cheap, steady power from REA lines, and most fell into disuse.

Several European countries built enormous wind generators. During the 1950's and 1960's, the French built advance-design

100-kW to 300-kW units. The Germans built 100-kW wind generators to provide extra power for their utility lines. But because of stiff competition from cheap fossil-fuel generators, these experimental machines were eventually decommissioned.

One of the most memorable wind machines was the Smith-Putnam machine built near Rutland, Vermont, during the 1940's. This huge machine with 175-foot blades was designed to deliver 1250 kW to the Vermont power grid. For a short time it delivered 1500 kW. But wartime material shortages and lack of money brought an end to this project after high winds broke one of the two 8-ton blades.

It's easy to talk of significant design improvements and new uses for wind power, but very few of the mistakes and disasters that occurred along the way appear in the historical record. Rather, the lessons learned from these errors have been incorporated into the evolving designs. No doubt there is a principle of natural selection surreptitiously at work in the area of windmill design. Today, we have cheap electronic calculators, pencil erasers, and a small number of books available to help us discover our mistakes *before* we erect our machines.

If you want to design and construct a successful wind power system, you should be aware of the valuable lessons learned from past successes and failures. The experience gleaned from scratched knuckles, broken wrenches, holes drilled in the wrong place, and toppled towers is contained in

Rated at 1250 kilowatts, the giant Smith-Putnam Wind Generator produced electricity for the Vermont power grid during World War II.

Sail-wing windmill at the New Alchemy Institute in Woods Hole, Massachusetts. In this aquaculture experiment, the wind machine pumps the water needed by a greenhouse ecosystem.

these pages. But before everything presented here hits home completely, you, too, must partake of the joys and sorrows of the mechanic, the technician and the engineer.

The Uses of Wind Power

Today, wind power can be harnessed to provide some or all of the power for many useful tasks such as pumping water, generating electricity, and heating a house or barn. Let's examine a few of these more closely.

Pumping water is a primary use of wind power. Daniel Halliday and others began manufacturing multibladed windmills for this

purpose in the mid-nineteenth century. Halliday's work coincided with advancements in the iron water-pump industry. Soon the combination of wind machines and iron water-pumps made it possible to pump deep wells and provide the water for steam locomotives chugging across the North American plains. The demand for wind-powered deep-well pumps created a booming turn-of-the-century wind power industry. Sears sold those machines for about 15 dollars, 25 for a tower.

The wind has also been harnessed to provide mechanical power for grain grinding, sawmill operation and even driving a washing machine. While I don't envision next year's Kenmore washing machine to come complete with a tower, blades and driveshaft in lieu of an electric cord and plug, mechanical power from a wind machine can prove useful.

Electricity can power just about anything, though, and its generation from wind power seems to grab the lion's share of attention. You can pump water, run washing machines, grind grain, heat houses and read books with electric power. As soon as electric generators from old cars became available, farmers started building "light plants," or homemade wind generators. Such early mechanics magazines as *Popular Science* and *Modern Mechanix* showed how to convert water-pumping windmills into wind-chargers using junked generators, bicycle chains and the family wind machine.

Many Midwest farmers already had gasoline or kerosene generators to charge their batteries, and the addition of wind power helped reduce fuel costs and wear-and-tear on generators. Out of all this backyard activity grew the pre-REA windcharger industry. Some half-million wind systems once existed in the United States alone, but it's not clear from historical records whether this number includes the water pumpers along with the windchargers.

Farmers used wind-generated electricity to power a radio, one or two lights for reading, eventually an electric refrigerator or a wringer washing machine, and not much else. Electric irons for pressing clothes, electric shavers, and other gadgets built to run on direct current appeared, but most of these proved unrealistic uses for wind-generated electric power. In fact, they may have contributed to the demise of wind electricity when rural electrification began. Electric appliances performed much better on an REA line, which wasn't subject to dead batteries. "Let's go over to the Joneses, Pa. They got one of them new power lines. Maybell says her refrigerator don't defrost no more!"

Rural electrification put most windchargers out of business. In the Midwest you can drive for miles on an empty dirt road, following a long electric power line to only one, or perhaps two, homes at the end of the road. Leave one road and follow the next. It's the same story. REA lines were installed and wind generators came down. Sears catalogs touted all the marvelous gadgets one could

buy and plug into the newly installed power line.

Electric stoves, hot curlers, electric air conditioners, two or more TV sets—these aren't very realistic loads to place on a wind-charged battery. However, wind power *can* contribute to the operation of these devices, especially if grid power is already doing part of the job. With such *cogeneration* (wind power used together with grid power) the more wind power available, the less grid power needed.

In another application, wind power can provide heat for warming households, dairy barn hot water, or just about anything else for which heat is used as long as the heat is not needed in a carefully controlled amount. This wind heating concept is called the *wind furnace,* and it's one of our most useful applications of wind power. Wind furnaces can use wind-generated electricity to produce the heat, or they can convert mechanical power into heat directly.

Energy Budgets

Wind machine design must begin with a realistic assessment of energy needs and available wind resources. When confronted by inexperienced people observing my wind machine, I'm most often asked, "Will it power my house?" Taking this question to its most outrageous extreme, I'm often tempted to reply, "Just how fast would you like your house to go?" But usually, I just ask, "How much power do you need at your house?"

A windcharger of the 1930's. Hundreds of thousands of midwestern American farm homes were powered by the wind before utility lines were installed by the REA.

Energy and Power

A clear distinction must be made between energy and power—two different but closely related quantities. Briefly, power is the rate at which energy is extracted, harnessed, converted or consumed. It equals the amount of energy per unit time, or

$$Power = \frac{Energy}{Time} .$$

An equivalent relation between these three entities is:

$$Energy = Power \times Time .$$

The amount of energy extracted or consumed is therefore proportional to the elapsed time. For example, a typical light bulb draws 100 watts of electrical power. One watt (1 W) is the basic unit of power in the metric system. Leave the light bulb on for two hours, and it will consume 200 watt-hours (100 watts times 2 hours equals 200 watt-hours, or 200 Wh). Leave it on for ten hours and it consumes 1,000 watt-hours, or one kilowatt-hour (1 kWh), the more familiar metric unit of energy.

In the English system, energy is measured in foot-pounds, British Thermal Units, and a host of other units that don't concern us here. One foot-pound (1 ft-lb) is the amount of mechanical energy needed to raise one pound one foot high. One British Thermal Unit (1 Btu) is the amount of thermal energy needed to heat one pound of water 1°F. Power is most often measured in horsepower and in Btu per hour. One horsepower (1 hp) is the power required to raise a 550-pound weight one foot in one second:

$$1 \ horsepower = 550 \ \frac{foot\text{-}pounds}{second} .$$

Note that the units of power are expressed in units of energy per time, as one would expect.

Conversions between metric and English units require that you know a few conversion factors. For example, one horsepower equals 746 watts, and one kilowatt-hour is equal to 3,413 Btu. Thus,

$$100 \ watts = \frac{100}{746} \ hp = 0.134 \ hp ;$$

$$10,000 \ Btu = \frac{10,000}{3,413} \ kWh = 2.93 \ kWh .$$

Many more conversion factors are presented in Appendix 1.2, along with a brief explanation of how to use them.

Rarely is it ever a simple task to estimate the wind energy available at a particular site; the windspeed is constantly changing. During one minute, 300 watts of power may be generated by a windmill, or 300 watt-minutes of energy (which equals 5 Wh). During the next minute the wind may die and you get absolutely no energy or power from the machine. The power output is constantly changing with the windspeed, and the accumulated wind energy is increasing with time. The wind energy extracted by the machine is the summation or total of all the minute-by-minute (or whatever other time interval you care to use) energy contributions. For example, if there are 30 minutes during a particular hour when the windmill is generating 5 Wh and another 30 minutes when there is no energy generated, then the machine generates 150 Wh (5 × 30 = 150) of wind energy. If there are 24 such hours a day, then 3600 Wh or 3.6 kWh are generated that day.

Blank stares, mumbled confusion, sometimes ignorant silence follows. Then, "Well, will it power the average house?"

Apparently many people would like to install a $1,000 wind machine and "switch off" good old Edison. This is a fantasy. It might be reasonable to use the wind to power your house if old Edison is a $30,000 power line away from your new country home, but most end uses for wind power will be somewhat less extravagant.

A successful wind power system begins with a good understanding of the intended application. For example, should you decide that water pumping is the planned use, you must determine how high the water must be lifted and how fast the water must flow to suit your needs. The force of the wind flowing through the blades of a windmill acts on a water pump to lift water. The weight of the water being lifted and the speed at which the water flows determine the power that must be delivered to the pump system. A deeper well means a heavier load of water; speeding up the flow means more water to be lifted per second. They both mean more power required to do the job, or a larger *load*.

This concept of load is crucial to the understanding of wind power. Imagine that instead of using a windmill you are tugging on a rope to lift a bucket of water from the well. This lifting creates a load on your body. Your metabolic process must convert stored chemical energy to mechanical energy; the *rate* at which your body expends this mechanical energy is the *power* you are producing. The weight (in pounds) of the bucket and the rate (in feet per second) at which you are lifting it combine to define the power (in foot-pounds per second) produced. How long you continue to produce that power determines the total amount of mechanical energy you have produced.

The kind of application you have in mind pretty clearly defines the load you will place on your wind power system. Knowing something about that load will allow you to plan an *energy budget.* "What the #$%&," you ask, "is an energy budget?" Let's explain it with an analogy. When you collect your paycheck, you have a fixed amount of money to spend. You probably have a budget that allocates portions of your money to each of the several bills you need to pay. Hopefully, something is left over for savings, a few beers, or whatever else you fancy. Energy should be managed the same way, and if you live with wind power for very long, you'll soon set up an energy budget.

Setting up an energy budget involves estimating, calculating, or actually measuring the energy you need for the specific tasks you have in mind. If you plan to run some electric lights, you must estimate how many, how long, and at what wattage. If you plan to run a radio for three hours each evening, you'll have to add that amount of electrical energy to your budget. If you want to pump water, you should start with estimates of how much water you need per day and calculate how much energy is required

to pump that much water from your well into the storage tank.

Wind furnaces require that you calculate the amount of heat needed. In some cases, you only need to calculate the heat needed to replace the heat lost from your house when it's windy. Such a system works only when it's needed. Your energy budget will now be in heat units—probably British Thermal Units (Btu), which can easily be converted to horsepower-hours (hph) or kilowatt hours (kWh)—energy units more familiar to wind machine designers.

Your utility bills and the equipment you already own will help define your energy needs. For example, average electrical energy consumption in U.S. residences is around 750 kWh per month, or about one kilowatt-hour per hour. Or, more specifically, most residential well pumps are rated at one to three horsepower. You can easily determine how long your pump runs and arrive at the total energy required per day, per week, or per month. In short, you really need to get a handle on your energy needs before you can proceed to the design of a wind power system.

Wind Resources

You will also need to determine your energy paycheck. There are two possible approaches: (1) Go to the site where you intend to install the wind machine and analyze the wind resource, or (2) go search-

Wind Energy and Wind Power

Energy and power are derived from the wind by making use of the force it exerts on solid objects, pushing them along. Buildings designed to stand still against this force extract very little energy from the wind. But windmill blades are designed to move in response to this force, and wind machines can extract a substantial portion of the energy and power available.

The wind energy available in a unit volume (one cubic foot or one cubic meter) of air depends only upon the air density ρ (Greek "rho") and the instantaneous windspeed V. This "kinetic energy" of the air in motion is given by the formula:

$$\frac{kinetic\ energy}{unit\ volume} = \frac{1}{2} \times \rho \times V^2 \ .$$

To find the kinetic energy in a particular volume of air, you just multiply by that volume. The volume of air that passes through an imaginary surface—say the disk swept out by a horizontal-axis windmill—oriented at right angles to the wind direction is equal to:

$$Volume = A \times V \times t \ ,$$

where t is the elapsed time (in seconds) and A is the area (in square feet or square meters) of the surface in question. Thus, the wind energy that flows through the surface during time t is just:

$$Available\ Energy = \frac{1}{2} \times \rho \times V^3 \times A \times t \ .$$

Wind power is the amount of energy which flows through the surface per unit time, and is calculated by dividing the wind energy by the elapsed time t. Thus, the wind power available under the same conditions as above is given by the formula:

$$Available\ Power = \frac{1}{2} \times \rho \times V^3 \times A \ .$$

Both energy and power are proportional to the cube of the windspeed.

If all the available wind power working against a windmill rotor could be harvested by the moving blades, this formula could be used directly to calculate the power extracted. But getting such an output would require that you stop the wind dead in its tracks and extract every last erg of its kinetic energy. This is an impossible task. Some non-zero windspeed must occur downstream of the blades to carry away the incoming air, which would otherwise pile up. Under ideal conditions, the maximum power that can be extracted from the wind is only 59.3 percent of the power available, or

$$Maximum\ Power = \frac{0.593}{2} \times \rho \times V^3 \times A \ .$$

In practice, a wind machine extracts substantially less power than this maximum. For example, the windmill rotor itself may capture only 70 percent of maximum power. Bearings will lose another few percent to friction; generators, gears, and other rotating machinery can lose half of whatever power remains. Pushrods, wires, batteries and monitoring devices will lose still more. The overall "system" efficiency of the entire wind machine is the fraction of the wind power available that is actually delivered to a load or to a storage device:

$$Efficiency = \frac{Power\ Delivered}{Available\ Power} \ .$$

Thus, the power extracted by a particular wind machine with system efficiency E is given by the formula:

$$Extracted\ Power = \frac{1}{2} \times \rho \times V^3 \times A \times E \ .$$

The final output of a wind machine is greatly reduced from the power that is really available in the wind. In practice, values of E commonly range from 0.10 to 0.50, although higher and lower values are possible.

One more factor is needed before the formula above can be used in your calculations—a conversion factor that makes the answer come out in the appropriate power units, whether metric or English. The chart presented here gives the necessary values of this factor, K, which adjusts the calculation so that the result is expressed in horsepower, watts or kilowatts. The final formula combines everything so far presented:

$$Power = \frac{1}{2} \times \rho \times V^3 \times A \times E \times K \ (Eq.\ 1) \ .$$

This is a very important formula—perhaps the most important in this book. The remaining chapters help you to obtain the various factors needed to use this formula in your design procedure.

Wind Power Conversion Factors			
For:	Values of K to get wind power in:		
	Horsepower	Watts	Kilowatts
Windspeed in mph Rotor area in ft²	0.00578	4.31	0.00431
Windspeed in ft/sec Rotor area in ft²	0.697	520	0.520
Windspeed in meters/sec Rotor area in (meters)²	0.00183	1.36	0.00136

ing for the best wind site you can find. The former approach is more direct. You own some property, there is only one clear spot, and that's where the tower will be planted, along with your hopes for a successful project. The latter approach offers more avenues for refinements and better chances for success. In any event, the larger the paycheck, the less strain on your energy budget—the smaller and less efficient a windmill needs to be.

In either approach you need to measure, estimate or predict how much wind you can expect at your chosen site. I've talked with folks who claimed to be wind witchers, possessing the ability to use a wet index finger and predict the windspeed with great accuracy. I've talked with people who installed wind generators back in the days before rural electrification. Most of these machines were installed in areas now known to be quite windy, with average windspeeds of 14 to 16 miles per hour (mph). When asked, these folks almost always guessed that the wind averaged 30 mph or more. Those old windchargers were installed haphazardly. "Heck, anywhere you stick one, it will work just fine."

Once you have established an energy budget, you have effectively established a standard for the performance of your wind power system. That puts you in a different league than the pre-REA folks. Your system will be good if it meets the energy budget. Theirs was good because it was all they had. Your site analysis should be careful and conservative. If it is, the wind system you plan will probably serve its purpose. If not, you'll have to be happy with what you get, just like the folks before the REA. As you gain familiarity with your system, you might learn how to save a little power for later. Or maybe you'll build a larger wind machine because you like wind power so much.

From an engineering standpoint, the available wind power is proportional to the cube of the windspeed. Put another way, if the windspeed doubles, you can get eight times the wind power from it—unless the tower collapses. If you are off by a factor of two on your windspeed estimate, you'll be off by a factor of eight on the power you think is coming to you. Remember those farmers who guessed their average windspeed to be 30 mph, although it's been measured at around 15 mph. A factor of two.

Folks who have lived in an area for a long time can usually tell you what seasons are windy and the direction of the wind during those seasons. But they're not very good at estimating windspeeds. You'll have to measure the windspeed yourself, or use some accurate methods to estimate it.

System Design

Once you have established your energy budget and collected adequate wind resource data, you can begin the task of system design. Whether you intend to design and build the windmill yourself, assemble

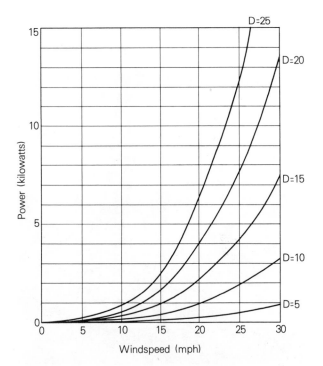

Wind power extracted by typical rotors with different diameters (D, in feet). An overall efficiency of 30 percent was used to calculate these curves.

An American Farm Windmill and water storage tank. Systems like this pumped water for the first railroads to cross the United States; farmers still use them throughout the world.

one from a kit, or buy one off the shelf, these are necessary preliminaries. Too often, designers disregard energy budget and site analysis. Instead, they make arm-waving assumptions on the use of the wind machine, where it might be installed, and use other criteria like cost as the design goal. Some of the best wind machines on today's market have been designed this way, but it's very important to realize that most manufacturers of mass-produced wind machines have left the tasks of energy budget and site analysis up to the buyer. The information provided in later chapters will help you make these analyses *before* you select a factory-built wind machine suited to your needs or set about designing and building your own system.

A thorough energy budget should tell you something about the time of day, week, or month when you need certain amounts of energy and power. For example, if everybody in your house rises at the same time each morning, your electric lights, hot curlers, coffee pot, TV set, toaster, stove, hot water heater and room heaters probably become active all at once. An enormous surge in electricity use occurs—the same problem Edison has. Their generators idle along all night doing virtually nothing until everybody wakes up at once. If you had a wind-electric system, it would be really nice if the wind at your site were strongest when the loads on the system were the greatest. Chances are very great, however, that your loads don't coincide with the wind. Hence, you can either try to synchronize your loads with the wind or store wind-generated energy until you need it.

The methods of energy storage are legion, but only a few are practical. If wind is being used to pump water, your energy storage might be a familiar old redwood water tank. Electrical storage has traditionally taken the form of batteries—still the most reasonable means of storage in many installations. Cogeneration allows you to send wind-generated electricity out to utility lines (running the meter backward) when you don't need it all. In effect, old Edison becomes the energy storage for the wind system.

There are a number of exotic ways one might choose to store excess wind energy. You might dynamite an enormous mine under your house and pump it up with air from a wind-powered compressor. This compressed air can then power a small generator size for your loads, as well as provide aeration for the tropical fish tank. If you own enough land, you can bulldoze a large lake and pump water up to it with wind power. A small hydroelectric turbine will produce electricity as you need it. In fact, you might sink two telephone poles out in the yard and use a wind-powered motor to power a hoist, lifting a '56 Oldsmobile up to 100 feet. As it descends, the motor that lifted it becomes a generator. Such a mechanism could provide you with 500 watts of electricity for about 15 minutes—maybe enough to burn the toast! There are as many possibilities for energy storage as there are crackpot inventors around, and some of these possibilities are just as crazy.

Some systems provide energy storage as an inherent part of the design. The wind furnace, where wind power is being used exclusively to produce heat, is a good example. Heat storage is energy storage—you may store energy by heating water, rocks or a large building with the excess heat. But probably you will be heating with wind power when you need heat the most. Wind chill can draw heat from a house much more quickly than occurs under no-wind conditions. So little, if any, storage would be necessary. But for most applications, some energy storage is mandatory.

Wind system design is a process of balancing energy needs against wind energy availability. Besides picking a good site and buying or building the right wind machine, you have to select a suitable storage system, plan all wiring or plumbing, build a tower, support it with guy wires, and get building permits and neighbor approval. This design process can be conveniently summarized in the accompanying flow chart. To follow this chart, you start where it seems appropriate and follow the arrows, completing the task in each box before proceeding to the next. This book is organized to help you use this flow chart in your design process. Whether you intend to design and build the entire system or just assemble it from factory-built parts, this book will help you achieve a wind power system worthy of your efforts.

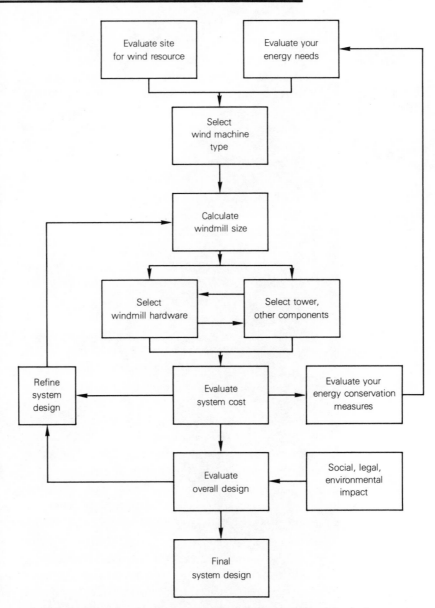

Steps involved in planning and designing a successful wind power system.

2
Wind Power Systems

Some Practical Applications of Wind Machines

Wind power has a surprisingly wide range of applications. Dairies, apple growers, hog farms, schools, electronic telemetry users, and many rural home owners are examining their resources and needs and trying to balance them in their wind system designs. But few, if any, of these systems will ever supply 100 percent of user energy needs. Harnessing the wind to supply anywhere from 25 to 75 percent of designated energy needs is a more reasonable expectation.

Different applications require different wind system geometries (including design, materials, and implementation). For example, to pump water or compress air you'll need lots of torque produced by a rotor with high total blade surface area operating at low rpm. If you want to generate electricity, however, you'll need just the opposite—a low-torque system with low blade surface area that spins very fast.

Remember that a "mix" of energy sources is available to the modern wind-energy user. Prior to rural electrification, farmers charged their dead batteries with a gas or kerosene generator. Or they carted them to town to be recharged at a repair shop while an afternoon was spent "out on the town." Today's energy mix means that a wind-energy user can supplement wind power with many other sources. Jumper cables from the family auto will boost a 12-volt battery bank. A gasoline, kerosene, or diesel generator or even the utility lines are typical back-up energy sources for today's wind systems.

As you study and use the information in this book, you'll discover the many ways in which the desired application determines the system design. In this chapter, various wind systems are introduced along with some cogent experiences. First we'll look at a wind-powered water pump that uses fabric sails—not unlike thousands of Cretan wind machines that have pumped water for centuries.

A Sail-Wing Water-Pumper

John Welles is an inventive tinkerer who has installed a number of wind-powered water-pumping systems around Northern California. In 1977, John loaded a "sail-wing" wind machine into the back of his Volkswagen Beetle and delivered it to my ranch. We set it up in only two days.

Water is available at this site from an artesian spring close to the surface; it does not need to be raised from deep in the ground. Wind power supplies the pressure needed to transport that water along 300 feet of plastic pipe and up 2 feet of elevation to a stock-water tank used as a reservoir for a small goat dairy and a trickle irrigation system for the organic gardens at the ranch. Much of this water pressure occurs because friction forces work against water flowing swiftly through a long water pipe. This friction back-pressure is referred to as the *friction head.* The total power the wind pump must generate is based on the total height water must be lifted (in this case only a few feet),

Installation of a sail-wing rotor at the Park ranch in California. This 16-foot diameter rotor was eventually replaced by a larger version.

the friction head, and the rate at which the water flows.

The design of this sail-wing windmill evolved from a notion that only hardware store components and materials would be used. The entire rotor, support structure, and pump are made from iron and plastic plumbing components. The sail spars are made from electrical conduit tubing. And the "tower" is a redwood fence post with clothesline-cable guy wires.

Originally the three sails spanned a diameter of 16 feet, but they now span 20 feet. In water pumping systems, the wind rotor (blades, hub and powershaft) must start turning under wind power against a heavy load of water. Rotor design for this type of load usually calls for a fairly large total blade surface area. This is why the familiar farm water-pumper has so many blades—it needs high starting torque. The term for the ratio of blade area to frontal area is *solidity*. The more blade surface area, the more "solid" the frontal area. The sail-wing machine at my ranch has few blades—or a small blade area relative to the large frontal area of the entire three-bladed rotor. Hence, it doesn't have much starting torque; fortunately, it doesn't need much because it only lifts water a few feet. With the 16-foot diameter rotor, the machine began pumping when the windspeed reached about 10 mph. The larger diameter rotor lowered this "cut-in windspeed" to about 7 mph. More blade area means higher starting torque and a lower cut-in windspeed. Extra area can be added by sewing wider sails or by adding more sails similar to those already installed. More blade area, however, means greater loads on the guy wires or cables that support the tower, because there is more surface area for the wind to push against.

Sail shape also plays an important role in determining the ability of the wind machine to pump water. These sails billow and flap about in the wind a bit too much. It's probably impossible to achieve perfection in sail design, but government and private research programs are exploring windmill sail designs to improve performance.

The sails on the rig at my ranch are sewn from sailboat-quality dacron cloth, although canvas or other materials could be used. Dacron is lightweight and very strong. It's also one of the few fabrics that can last a few years in an extreme outdoor climate. Freezing weather and strong sunlight combine to destroy the fabric eventually, but it lasts long enough to make a sail machine worthwhile. Screendoor springs connected to the sails hold them taut for normal operation. These springs stretch under sail loads imposed by high winds, allowing the sails to "luff," or flap, out of the wind. This simple "governor" protects the sails from damage. Because the governor lessens the sail loads, tension loads in the guy wires that support the tower remain low enough during high winds that short stakes such as goat-tether stakes are adequate for tie posts.

A simple crankshaft translates rotary blade motion into the up-down stroke needed

The sail-wing machine in action, pumping water from an artesian well close to ground level.

Sucker rod and water pump used with the sail-wing. The sucker rod pushes down on leather pump seals, driving a pulsing stream of water into the plastic pipe at rear.

to drive a water pump. This pump, a piston 3 inches in diameter with leather pump seals, is a "single-acting" pump. That is, it pumps only on the down stroke. This pump pushes water into the pipe in pulsing streams. However, peak water flow rates can be too fast for efficient operation. The faster that water flows in a pipe, the higher the friction and the resulting back-pressure against the pump. By installing a surge chamber (a simple tank with trapped air) in the water line near the pump, we created an air space where strong water pressure pulses could expand and slow down the peak flow rate. More continuous water flow resulted, and more water was pumped because of improved pump performance.

This sail-wing water-pumper has worked well for over a year, with only minor problems associated with the realities of a do-it-yourself system. For example, the 300-foot plastic water pipe was not buried at first, and whenever a horse stepped on the flexible polyethylene, a water lock was created that stopped the pump. Occasionally this occurred during high winds and broke a "sucker rod"—the long, thin tube that connects the crankshaft to the pump. Solutions included a stronger sucker rod and—finally—burying the water pipe.

Performance has been adequate. The storage tank requires about 400 gallons of pumped water per week. The wind resource at the site is minimal, about 8 mph annual average, but evening breezes of about 11 mph drive this sail machine long enough

to keep the stock full.

This sail-wing water-pumper can be readily adapted to many sites. With deeper water, however, more sail area would be required to provide the necessary starting torque. The reliability of such a machine is directly related to the amount of care you put into the project and the time you devote to solving its early problems.

The Old Farm Water-Pumper

A giant step up from a simple, homemade water-pumper is the old multiblade pumper still available today—new or recycled. Such machines have been in use since about 1860. Dozens of them can still be seen in windy areas. This machine is often called "Ole Reliable." It just keeps on pumping even though the owner may not perform any maintenance for several years. One farmer I talked with hadn't checked the oil level in his machine for 25 years! It's not surprising that most of the problems with these machines are caused by lack of owner care.

The process of erecting such a machine starts with collecting all the pieces necessary for complete installation. When you buy an old machine, and occasionally when you buy a new one, some parts are likely to be missing. Don't try to fake it. You need *all* the parts. Leave one bolt out and the whole contraption is apt to end up on the ground! Laying the foundation is followed by tower erection. Here, there are two schools of thought in the wind industry. In the first, you assemble the tower on the ground and tilt it up. In the second, you assemble the tower from the ground up. Both ways work, and the method you choose will probably depend on the site. If you have lots of room for the machinery necessary to tilt up a tower, that's the easier way to go. If not, you can erect the tower straight up, but it'll be more difficult because of having to handle unsupported tower parts atop a partially assembled tower. Either way, the tower must be absolutely vertical. Otherwise, the windmill, which rotates about a *lolly,* or *yaw,* axis (the vertical shaft that allows the windmill to change its direction), will not aim properly into the wind.

There are several ways to hoist a wind machine aloft. You can do it by using a block-and-tackle at the top of the tower. You can carry it up piece by piece—not much fun, but a lot of exercise. Or you can lift it up with a large crane or hoist.

Next you'll need to install the sucker rod, which drives the well pump in its up-down motion. Traditionally sucker rods have been made of steel or wood. Wind blows against these long, slender rods and causes them to bend. When the cyclic, tension-compression loads of the pump and wind rotor are added, they can buckle or break.

Pumps are pistons that slide along inside metal or plastic pipe, pulling water in through a one-way "foot valve" and pushing it through a pipe to the surface. Usually pumps are installed deep in the ground. Mine is 60 feet

The reconditioned water-pumper at the Park ranch was carried aloft piece by piece and reassembled at the top of the tower.

With a diameter of 8 feet, this wind machine can lift about 500 gallons of water per hour in a 15 mph wind.

The multiblade water-pumper at my ranch is 8 feet in diameter. In a 15 mph wind, it can lift about 500 gallons per hour (gph) from 60 feet below ground level. The well tested at 600 gph when it was bored, so this water-pumper probably won't run the well dry. Its task is to fill two 400-gallon tanks—one tank every week and the other once per month. Hence, it needs to pump about 2,000 gallons per month. Some months are windy enough; others are not. Thus, a gasoline-powered well pump was installed in the same bore hole as the wind-powered pump. This back-up pump can be started whenever a storage tank runs dry and no wind is blowing.

This installation illustrates all of the elements of a complete wind system. It has a wind energy converter—the "windwheel," or rotor—that converts the kinetic energy, or force, of the wind into rotary torque in a powershaft. A crankshaft converts that rotary motion into up-down, or reciprocating, pump motion. A sucker rod transfers this up-down motion down the tower and into the ground to the pump. Pipes carry the pumped water from the well to two storage tanks. And a back-up gasoline-powered pump provides power when water needs exceed the capabilities of the wind resource.

down; some are as deep as several thousand feet. Unfortunately, dirt plays havoc with pump seals and wears them out. As you might suspect, the major maintenance required by water-pumping windmills is pump overhaul and sucker rod replacement. One Montana farmer who owns several hundred of these machines has a work crew whose only job is to overhaul a different pump each week.

Wind-Electric Systems

Apart from water-pumping windmills (and European grain-grinders), windchargers were the only other large-scale application of wind power. A typical windcharger/battery system of the 1930's consisted of a three-bladed wind generator—mounted atop a 50-foot steel tower—that charged a bank of nickel-iron or lead-acid batteries. Often the batteries were also charged by a back-up generator powered by gasoline or kerosene.

Early loads included radio sets and lights—loads operating at 32 volts. Just before the onset of rural electrification, 110-volt wind systems became plentiful. As demand for convenience items rose, farmers added more batteries to their storage banks and occasionally upgraded to larger wind generators. They added 32-volt direct current (DC) wringer washers, laundry irons, refrigerators, freezers, and electric razors to the list of loads. The use of wind-generated power grew at individual installations to such an extent that farmers were "ready" for rural electrification when it came along offering "all the power you want."

Perhaps the most successful wind generator built during this era was the Jacobs. During the late 1930's this machine, and company, became the sole proprietorship of the now famous Marcellus Jacobs, a creative salesman who was able to market the work of his inventive brother. Indeed, many owners of other wind machines eventually switched to a Jacobs.

Several Jacobs models were available, ranging from about 1,800 watts to 3,000 watts output. A typical 3,000-watt (3 kW) machine had three wooden blades and a

Close-up view of an old Jacobs wind generator. With wooden blades and a rotor diameter of 14 feet, this model generated up to 3,000 watts of electric power.

rotor diameter of about 14 feet. The three-bladed rotor and airfoil blade design provided the low solidity and high rpm required to generate electricity. Its maximum output power occurred at a windspeed of 27 mph. These machines were installed along the East Coast of the United States and in large numbers in the American Midwest—the Dakotas, Montana, Wyoming, and southward. After REA penetration, many of them were transplanted to parts of southern

A recycled Jacobs in action at Windworks in Mukwonago, Wisconsin. Many of these machines have been reconditioned and are now producing electricity at remote sites.

Fiberglass-bladed rotor being tested on the back of a pickup truck. Wind machines are usually tested, often to complete destruction, before final designs are ready.

Canada not yet served by power lines. Some Montana installations were still in use as late as 1959. Rebuilt versions of this machine are available today, but the supply of rebuildable machines is dwindling rapidly.

Early wind machines, the Jacobs included, had problems that required solutions. Many machines flew apart as a result of stresses induced by centrifugal force. Governors of all types were invented, tried, and patented. Large wind companies bought out smaller ones to get patents they wanted. Generators, gear boxes, and blades of all sorts were developed. Some failed, some survived. While water-pumpers were being built much the same as their turn-of-the century predecessors, wind generators were modern, glamorous creations that served the needs of convenience. With rural electrification, however, convenience flourished, and wind-powered farm homes virtually disappeared.

Evolution of a Wind-Electric System

My own ventures into wind energy began with an avid desire to use surplus helicopter rotor blades for electricity production. In the late 1960's and early 1970's, there was no wind-energy literature available in the county library, so I set about deriving the various aerodynamic equations needed to design a good windmill.

My first blades were constructed with a 2-inch aluminum tube as the main load-carrying spar, with a fiberglass skin rivetted and epoxy-bonded in place around this spar. This skin was laminated with polyester resin over a shaped male plug mold and removed from the plug after the resin cured. Ribs were sawed from half-inch-thick plywood to close each end. The result was a 6-foot long blade that weighed 8 pounds.

But the first windmill, tested on the back of my pickup truck to simulate a wind tunnel, was a disaster. The fiberglass-bladed rotor would not spin fast enough to cause a generator—even with a speed-up transmission— to kick in and start charging the batteries. The blade design had too much *twist* (spiral turn). In a twisted blade, the airfoil at the root end (closest to the center of the rotor) points more into the wind than the airfoil at the tip of the blade. Because of the excessive twist in my first blades, the root end of each blade was acting like a giant air brake and preventing the rotor from reaching the necessary rpm.

The lessons learned from that first rotor design caused me to re-evaluate my blade-design equations. The new equations are the basis for many of the calculations discussed in this book; they correct the over-twist of my first blades and produce blade designs that are efficient and easy to build.

I also tried a different method of blade construction that eliminated the need to mix resin and laminate fiberglass. Because it offers a smoother surface and less air friction, I chose aluminum for the skin of my next blades. But unfortunately sheet aluminum

The second in a series of experimental rotors being readied for a test run. To lower the air friction, this rotor used sheet aluminium instead of fiberglass for the blade skins.

Drag brakes used as a governor on a small, fiberglass-bladed rotor. At high rotor rpm, these flaps extend out from the blade tips, slowing the rotation speed and protecting the rotor.

doesn't like to be shaped or curved in more than one direction at a time. Thus, the normal airfoil-shaped curvature of the blade threatened to eliminate blade twist because twisting an already curved skin would demand compound curvature—skin curved several ways at once. I could only twist my blade 6 degrees from end to end without buckling the skin. That's not very much. With a little more blade twist, the new rotor wouldn't have taken so long to start spinning.

Electric generators don't produce current until they are spinning at a high rpm. My first rotor developed strong starting torque because of the large amount of twist, but could not spin fast enough to power a generator. The second rotor had less twist and was built with a sheet aluminum skin that was much smoother than the earlier fiberglass skin. Because of the reduced twist, the second rotor was a bit slower to start spinning, but once it got going it began to spin really fast—as high as 400 rpm. The improved performance was a direct result of optimizing the blade design and lowering the surface friction.

High-speed operation requires the use of a good governor to keep the blade rpm within structural limits. The first governor I used was a combination of drag brakes and a flexible hub. The drag brakes, or flaps, were designed to extend from the blade tips whenever rotor rpm was high enough for spoiler weight (inertia plus centrifugal force) to overcome the tension of a prestretched spring inside the blade. And the flexible,

spring-loaded hub permitted the blades to form a cone in the downwind direction. This reduced the rotor frontal area, thereby reducing power and rotary speed. Proper governor design requires that all blades behave the same. All drag brakes should be interconnected so they operate together. If they don't, severe vibration will set in. In my system, severe vibration did set in, but the free-coning hub worked well and will be used again in future machines.

The next governor I tested used flyballs (lead weights) that were rigidly attached to each blade near the hub in such a way that they would move into the plane of rotation when the rpm became excessive. The blades would then feather—point directly into the wind—and rpm would decrease. This "flyball governor" worked fine and is still being used today. A certain amount of "tuning" is necessary; flyball weights and springs need to be changed around by trial and error for optimum performance. In some cases a dampener similar to an automotive shock absorber might be necessary to adjust the rate at which the flyball governor works. In any case, my experience with several dozen variations of blades, governors, and generators shows that a governor is always necessary to keep the rotor spinning within its intended operating range.

Rotors and generators need to be matched to each other. For example, a typical rotor might be designed to spin at 300 rpm in a 20 mph wind, but most generators need to spin much faster than 300 rpm;

hence, a speed-up transmission is usually needed—unless the generator is a low-speed, very heavy unit borrowed from an old Jacobs machine and specifically designed to spin at the same speed as the rotor.

My early machines used automotive alternators that needed 2,500 rpm to generate full output. The first transmission was a chain drive bought at a go-cart shop. Chains, sprockets and bearings of all sorts are readily available at such places or at tractor and agricultural machinery shops. However, chains tend to be noisy, short-lived, and require frequent service.

The next transmission I tried was a 3-inch wide toothed belt with a 3:1 speed-up ratio followed by a second, narrower belt with another 3:1 ratio for a 9:1 total speed-up. Thus, 300 rpm at the rotor resulted in 2,700 rpm at the alternator. This transmission worked well. However, it is complex, requiring careful alignment and tensioning of the belts. It is also very heavy (because of the steel pulleys) and expensive if bought new. The final transmission I used was a gear box—actually an industrial *speed reducer.* In a windmill we run it backwards as a speed increaser. Such gearboxes are available at bearing, chain, and pulley houses under any of several brand names. They are heavy, cheap and reliable. Best of all, you don't need to fiddle with them, just change the oil.

During any do-it-yourself or commercial project, I have noticed that the participants

Automotive alternator and belt transmission used with the second experimental rotor. Some transmission is usually needed to match rotor and generator rpm.

must make a conscious effort to combat a pernicious disease called "fire-em-up-itis." It's like a virus hidden within the human body, just waiting for the "nearly complete" syndrome to trigger it off. The result, in the case of a wind project, is usually, "Well, let's just let 'er spin a little bit." This happens before all the nuts and bolts are tightened, the batteries hooked up or something else that should have been done first. Often as not the disease results in broken blades, bent governor rods, or burnt-out generators (not to mention burnt-out people).

My first bout with this infection occurred when a friend and I had just installed our first flyball-governed, aluminum-bladed machine atop the tower and had to wait for a 12-volt battery to be charged (it registered only 10 volts on a voltmeter). We decided not to wait. The result was more waiting. With only 10 volts available, the alternator's voltage regulator seemed to be confused. The blades were spinning quite rapidly in a brisk wind, but no charging was taking place. At the time, we mistakenly decided the alternator was at fault. In reality, the voltage regulator was not designed to work with a dead battery.

Voltage regulators are not designed to work well with fully charged batteries, either. These transistorized controllers monitor battery voltage. If they detect that the battery is fully charged, they reduce the charging current from the generator. This has the effect of "clamping," or reducing, the output of a wind generator even when the wind is blowing quite hard. If your new system is trying to

charge a fully charged battery bank, fire-em-up-itis might lead you to suspect your new wind system is not "putting out what it should be." I've heard lots of folks complain about this.

Low-Voltage Technology

Jim Cullen of Laytonville, California, powers his home almost entirely on 12 volts of direct current. Numerous sources of electricity charge his batteries. On top of his house there's a small solar-cell panel. Nearby is a wind generator. And, when his batteries need an extra boost, jumper cables from his car add extra energy (presumably while warming the car up before a trip to town).

Far from the nearest utility line, Cullen's home sits on top of a mountain with a clear shot at the Pacific Ocean some 30 miles away. Daily average windspeeds range from 11 to 14 miles per hour—the minimum windspeed average needed for any type of successful wind-powered generating system. From the beginning, Cullen's concern was to devise a wind generator where little maintenance would be required, where most replacement parts could be obtained from local hardware outlets, and where cost would not be an obstacle. To meet these objectives, he enlisted the services of Clyde Davis, an old friend who also happened to be a consulting engineer. Davis designed two wind systems. One generates 160 watts of direct current in an 11–12 mph wind; the other

A close-up view of Cullen's 500-watt generator.

produces up to 500 watts DC under the same circumstances. Because the 500-watt wind generator is so light (only 125 pounds), an off-the-shelf tower was sufficient, and the entire cost for the system came to less than $2,000 (mid-1970's).

In conventional terms, 160 or even 500 watts isn't much power. But Cullen's house is anything but conventional. There are more than 1,500 square feet of living space filled with three television sets, approximately 20 fluorescent lights, a vacuum cleaner, washing machine, blender, mixer, water pump, a 40-watt per channel stereo system, and even a home computer. What makes all of this so unconventional is that Cullen has virtually eliminated the need for 110- or 220-volt AC electricity by converting his appliances (even the washing machine!) to operate directly from 12 volts DC. Table saws, drill presses and the like could be converted as well. Small appliances (e.g., the blender and mixer) that haven't been made to operate directly from 12 volts and for which there are no readily available replacement motors operate through an *inverter* that produces 110 volts AC from 12-volt batteries.

Cullen claims that a wind generator producing 500 watts can provide comfortable and reliable living at modest cost. Excess DC electricity produced in times of high wind can be stored effectively; high-voltage AC power cannot. Of course, such a system is no panacea. You have to become aware of the energy you use and plan accordingly. Things like heating, cooling

Jim Cullen uses both sun and wind to generate electricity at his Laytonville, California, home. A panel of solar cells atop the house and the 500-watt wind generator behind charge a bank of 12-volt batteries.

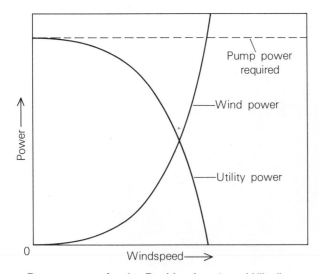

Power curves for the Bushland system. Utility lines supply whatever power cannot be provided by the wind.

The Darrieus rotor used for irrigation in Bushland, Texas.

and cooking are accomplished by appropriate alternatives that do not require electricity at all.

A Unique Darrieus Water-Pumper

At the other end of the spectrum from the sail-wing or farm water-pumpers is the Bushland, Texas, installation of a Darrieus "eggbeater" rotor tied mechanically to an electric 60-horsepower irrigation pump. This U.S. Department of Agriculture project at a windy location is examining practical approaches to water pumping with the wind. Since it is a relatively new research project, few performance data are available at this writing.

A Darrieus rotor has two or more curved, or bowed, airfoil blades that travel a circular path about a vertical power shaft at its center. Made with extruded aluminum, the streamlined airfoils have a rounded leading edge (the edge that cuts into the wind) and sharp trailing edge. The Darrieus rotor is linked to the pump powershaft by a clutch that allows the electric pump to turn freely but engages whenever the rotor is spinning rapidly enough. In periods of calm, an AC electric motor keeps the pump turning. The more wind power available, the less power drawn by the pump from the electric power grid.

This machine is also a *cogeneration system* that operates in parallel with the utility grid. It generates excess current that is delivered to the grid. In effect, the wind system becomes one of the many interconnected generators the grid uses for its power supply. At a certain windspeed, when more shaft power than the pump needs is available from the Darrieus rotor, the rotor overpowers the electric motor, trying to turn it faster than it was designed to turn. The motor then becomes an AC generator synchronized with the grid power. Should the electric power available actually exceed the demands at the site, the excess electrical energy is fed back into the grid, effectively "running the meter backwards." Thus, energy storage for this system is provided mostly by pumped water in a pond or a tank and occasionally by the grid power lines that "store" electricity sent backwards through the meter.

The unique advantage of the Bushland installation is that the electric motor actually serves three purposes—almost at once. First, the motor is the prime mover driving the well pump. Second, the motor is the governor for the wind rotor; it wants to turn only at an rpm determined by the AC frequency fed to it from the grid. By careful design, the rotor will never overpower the motor or cause it to overspeed by more than a few rpm. Finally, the motor is an electric generator whenever extra wind power is available.

Once perfected, this approach can be readily adapted to thousands of well pumps already in existence in the windiest parts of the United States and similar locales throughout the world. Minor modifications to these pumps will permit the addition of wind power,

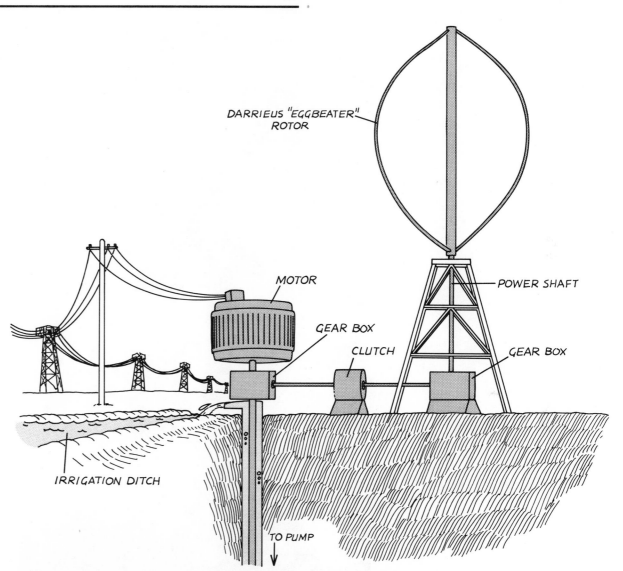

DARRIEUS "EGGBEATER" ROTOR

MOTOR

GEAR BOX

CLUTCH

POWER SHAFT

GEAR BOX

IRRIGATION DITCH

TO PUMP

In the novel approach being tested at Bushland by the USDA, both wind power and an electric motor are used simultaneously to power the well pump. If excess wind power is available, the system delivers it to the utility lines.

thereby reducing the amount of fuel needed to drive irrigation systems.

The Tvind System

With Denmark's vigorous history of wind power, it's not surprising that a significant wind system with cogeneration capabilities has been built by students and faculty at the Tvind School on the Jutland peninsula. This is the Tvind machine, rated at 2 million watts (2 megawatts) in a windspeed of 33 mph.

The fiberglass blades on this machine span a diameter of just over 175 feet, making it the largest wind machine in history at the time of this writing. Before the Tvind machine, the largest was the Smith-Putnam machine mentioned in Chapter 1. That machine was rated at 1.25 megawatts. Each blade of the three-bladed Tvind machine weighs about 5 tons; the two stainless steel blades of the Smith-Putnam each weighed about 8 tons.

The Tvind machine was designed to supply the school with its electrical and heating needs and to supplement the grid lines through use of a *synchronous inverter.* Alternating current from the generator is rectified to direct current then reconverted to grid-synchronized AC and fed to the school's electric system. The synchronous inverter is rated at 500 kW, while the school demands up to 1,500 kW. The balance between 500 kW of inverted power and 2,000 kW available from high winds will be used to heat water. The excess electricity is

The 2000-kW wind machine at the Tvind School in Denmark. This machine supplies both heat and electricity to the school.

fed to coils immersed in a hot-water storage tank. This feature makes the Tvind machine the world's largest wind furnace!

University of Massachusetts Wind Furnace

The wind furnace is an application of wind power that is finding widespread acceptance. Heat energy is needed to raise the temperature of water for agricultural and industrial applications ranging from dairy sterilization to washing laundry. It's also needed to warm animal barns, greenhouses, and homes. Happily, most of these climate control applications require more heat when it is windy than when it is not.

In a wind furnace, wind energy is converted into mechanical power in the rotor powershaft. This rotary power may then be used to make heat from friction (e.g., splashing water with paddles driven by the rotor power shaft) or from the generation of electricity that can power electrical resistance heaters.

Under the direction of Professor William Heronemous, students at the University of Massachusetts have built a wind generator 33 feet in diameter mounted atop a 50-foot steel pole tower. The electrical output of this machine is used directly to warm water, which is then combined with heated water from solar collectors mounted on the south-facing wall of the research house. Concrete tanks in the basement of the house store

Tvind School with the 2000-kW machine in the background.

all this hot water.

The fiberglass blades are capable of producing 50 horsepower (37 kilowatts) in a 26-mph wind. Shaft power from the blades drives a generator through a truck differential and a chain drive. The generator is controlled by an electronic load controller that senses windspeed and rotor rpm and applies appropriate current to the field windings

Above: Close-up view of the UMass wind generator.

Right: The University of Massachusetts wind furnace and test home. Both the wind generator and the south-wall solar collectors warm water stored in a concrete tank inside this home.

of the generator. In this manner, only the exact amount of power available from the wind is drawn from the generator, thus preventing overloading or underloading of the blades.

Only minor mechanical problems have been encountered in the operation of this machine. Most of these required simple mechanical adjustments; all have been solved as part of an ongoing educational program at the University. With the computerized data acquisition systems they are using, the students will provide the wind industry with a detailed history of one type of wind furnace.

I have included several types of wind machines here, but certainly not all of them. A Savonius rotor, a panemone, or a wind-powered hydraulic pump system—all of these and other projects will be discussed in greater detail. Your task is to select the design most appropriate for your energy needs. Always remember that good wind machine design begins with an assessment of your energy needs and the wind resources available.

3
Wind Energy Resources

Understanding and Measuring
the Winds at a Site

The wind can be a fickle servant. It may not be available when you need it, and you can be overwhelmed by its abundance when you don't. To harness the wind you must become familiar with its moods and be able to select a site suitable for a wind machine — a site where strong, steady breezes blow most of the year. After choosing a favorable site, you'll want to know how much wind energy is actually available and when it blows the hardest.

Good wind system design begins with a realistic assessment of energy needs and available wind resources. By comparing the two, you can estimate the size of the windmill you'll need. Storage size depends on how long the winds are calm when you need their energy. This chapter examines some basic principles of wind resources and develops some tools that will help you assess the winds at your site. Statistical data compiled for many years at airports and weather stations can give you a general grasp of average wind behavior, but there is no substitute for getting out to your site and measuring the wind resource directly.

Global Wind Circulation

Wind energy is a form of solar energy. The winds alleviate atmospheric temperature and pressure differences caused by uneven solar heating of the earth's surface. While the sun heats air, water and land on one side of the earth, the other side is cooled by thermal radiation to deep space. Daily rotation of the earth spreads these heating and cooling cycles over its entire surface. Seasonal variations in this daily distribution of heat energy are caused by seasonal changes in the tilt of the earth's axis relative to the sun.

Much more solar energy is absorbed near the equator than at the poles. Warmer, lighter air rises at the equator and flows toward the poles, while cooler, heavier air returns from the poles to replace it. In the Northern Hemisphere, the earth's west-to-east rotation bends northward-flowing air eastward and southward-flowing air westward. By the time the northward-moving air has reached 30°N latitude, it is flowing almost due eastward. These are the "prevailing westerlies," so called because they come out of the west.

Air tends to pile up just north of 30°N latitude, causing a high pressure zone and mild climates in these latitudes. Some air flows southward out of this high pressure area and is deflected west by the earth's rotation, forming the "trade winds" used by sailors the world over. A similar effect leads to the "polar easterlies" above 50°N latitude. South of the equator, the earth's rotation bends southward-flowing air to the east and northward-flowing air to the west. A similar pattern of prevailing westerlies, trade winds, and polar easterlies exists in the Southern Hemisphere.

Not all of the earth's surfaces respond to solar heat in the same way. For example,

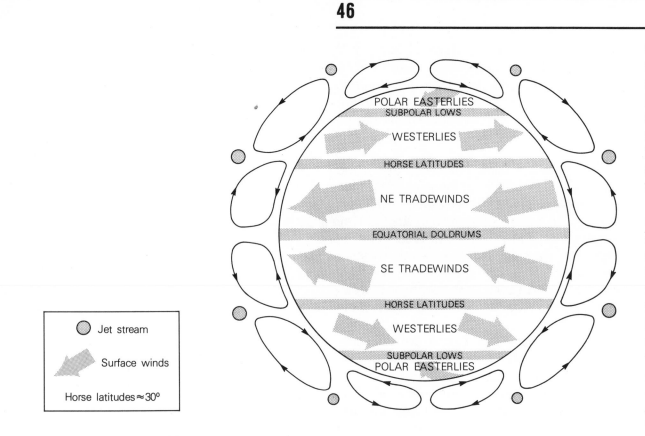

POLAR EASTERLIES
SUBPOLAR LOWS

WESTERLIES

HORSE LATITUDES

NE TRADEWINDS

EQUATORIAL DOLDRUMS

SE TRADEWINDS

HORSE LATITUDES

WESTERLIES

SUBPOLAR LOWS
POLAR EASTERLIES

○ Jet stream

⬟ Surface winds

Horse latitudes ≈ 30°

Global wind circulation patterns. Prevailing winds blow from the east in the tropics and from the west in the mid-latitudes.

wind-circulation patterns. When a warm air mass collides with a cold air mass, the resultant frontal activity generates most of the large-scale winds that drive the local winds at your windmill site. Imagine a globe covered by large bubbles that drift about bumping into each other. Whenever they bump, air is squeezed along—producing your local winds.

High and low pressure systems are associated with these air masses. In general, as air temperature increases, the density drops and so does the barometric pressure. The high pressure systems push their cooler air toward the warmer low pressure systems, trying to alleviate the pressure difference between the two. But the earth's rotation deflects these winds so that the air flows from high pressure to low along a *curved* path. In the Northern Hemisphere, the wind blows clockwise around a high pressure system and counterclockwise around a low. Together with the global circulation patterns mentioned before, the large-scale winds drive these same pressure systems around the earth's surface, bringing sunshine, cloudiness, rain, and more wind to the areas over which they pass. Accurate prediction of this detailed behavior is difficult, but each site has weather regularities that can be expressed as monthly and yearly averages. Some of these averages can prove very useful in describing the expected wind behavior at the site.

Large-scale winds generally dominate. But local winds often enhance or modify

an ocean will heat up more slowly than the adjacent land because water has a high heat capacity, or ability to store heat. Similarly, an ocean will cool down more slowly than the nearby land. These different heating and cooling rates create enormous air masses with the temperature and moisture characteristics of the underlying ocean or land mass. These air masses float along over the earth's surface, guided by the global

Air Temperature and Pressure

What we call atmospheric pressure is caused by the weight of the air above us. Travel to a mountaintop and lower atmospheric pressure occurs because you are nearer the top of the giant ocean of air surrounding the earth.

A barometer is commonly used to measure atmospheric pressure. A "falling barometer", or declining barometric pressure, has always been associated with an impending storm because a low pressure system is moving into the vicinity. A "rising barometer" heralds the approach of a high pressure system and fair weather. A typical winter distribution of atmospheric pressure is shown in the top map. The lines of equal pressure are called isobars, and the units used are millibars. At sea level, 1,013.2 millibars equals a pressure of 29.92 inches of mercury or 14.7 pounds per square inch (psi); all three values are equal to a "standard" atmospheric pressure. Inches of mercury are commonly used to measure pressure in the United States, and millibars is the metric unit of pressure used by scientists and airplane pilots.

The lower map shows average January temperature distributed over the globe. Note the correlation between temperature and pressure. Air temperature affects air density, which is related to weight or pressure of the atmosphere. Higher temperature air weighs less; air density therefore decreases with increasing temperature. A lighter air mass exerts less pressure at the bottom of the atmospheric ocean. So, as air temperature increases, its pressure decreases.

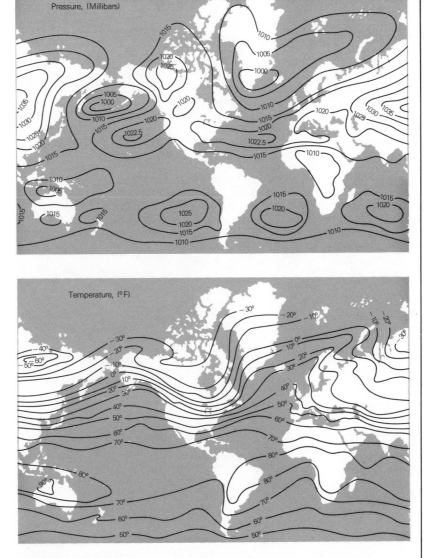

Global distributions of the average January air pressure and temperature.

Air Density

The density of any substance is a measure of how much of it can fit in a given volume. We usually want to know how much air, or how many molecules, interacts with a rotor as the wind passes through it. Kinetic energy in the wind is a function of the air mass and windspeed. Therefore, we need to know the mass density of air, ρ, to calculate wind energy and power. This density is based on the air temperature and altitude, usually referenced to sea level.

A "standard day" is an average day that scientists and engineers use for purposes of design. On a standard day, certain standard values can be used for temperature, pressure and density of the air (depending on altitude). Naturally, performance calculations on wind machines will vary somewhat from this standard, because a standard day is actually very rare. Summer days are generally hotter and winter days cooler. Standard conditions for air at sea level are, in English units:

Temperature = 59.9°F
Pressure = 14.7 psi
Density = 0.002378 slug/ft³.

Here, the "slug" is the English unit of mass, not a slimy animal resembling a snail. One slug weighs 32.2 pounds at sea level. To calculate the air density ρ at a specific windmill site, you need to correct for altitude and temperature differences from the standard case. Use the following formula:

$$\rho = C_A \times C_T \times 0.002378 \ slug/ft^3.$$
(Eq. 2)

The altitude correction factor C_A and the temperature correction factor C_T are taken from the two accompanying tables. For example, suppose that the average temperature at your site is 80°F, and the altitude is 2500 feet above sea level. Then $C_A = 0.912$ and $C_T = 0.963$ from these tables, and

$$\rho = 0.912 \times 0.963 \times 0.002378 \ slug/ft^3$$
$$= 0.002088 \ slug/ft^3,$$

or about 88 percent of the standard air density. More frequently, you will have to know these two correction factors at altitudes and air temperatures not listed directly in the tables. In such cases, just interpolate between the values given.

Altitude Correction Factor	
Altitude (feet)	C_A
0	1.000
2,500	0.912
5,000	0.832
7,500	0.756
10,000	0.687

Temperature Correction Factor	
Temperature (°F)	C_T
0	1.130
20	1.083
40	1.040
60	1.000
80	0.963
100	0.929

the large-scale winds and contribute some energy when none is available from large-scale winds. High pressure zones push air toward low pressure zones, and wind is created. The size and distribution of each pressure zone changes under the influence of the winds they create. Fast-moving air descends into valleys, blows through canyons, and sails over mountain peaks—redistributing warm and cold air masses. Because pressures and temperatures cause winds and are changed by those winds, it's impossible to predict the available wind power accurately by studying aviation weather forecast charts.

Because of its lower heat capacity, the temperature of a land mass rises and falls in response to solar energy and night-sky radiation more rapidly than the sea. Hence, sea water is cooler than the shore during the day, and warmer at night. Circulatory air flows, called "sea-land winds," are created by that temperature difference. You experience these winds as on-shore and off-shore breezes at the beach. During the daytime, especially in the afternoon, the warmer air over the land rises and the cooler air over the water flows in to replace it—creating an on-shore breeze. The reverse process creates off-shore breezes at night. Daytime sea breezes can be strong enough to be a source of wind energy; typically, they range from 8 to 16 mph. Nighttime breezes are slower, typically less than 5 mph.

Cool air next to mountain slopes is heated by exposure of its moisture to the morning sun and by the rising temperature of the ground below. As this high mountain air warms up, it rises; cooler air from the valley below is drawn upward along the slope. Central valley air flows toward the base of the mountain and up the slope. A "mountain-valley wind" is thus created by the temperature difference. At night, the process reverses itself; cooled by rapid radiation of heat to the dark night sky, high mountain air descends into the valley, gaining speed as it becomes heavier. Mountain-valley winds are generally thought to be too weak to be a source of wind energy. In some areas, however, this might not be the case.

Mountain breezes descending into a valley can be warmer or cooler than the valleys. Cool winds flow downhill under the influence of gravity, and tend to occur at night. Warmer winds gain heat energy by compression as they descend into valleys and may bring a local temperature rise of 30°F to the valley. This air is compressed as it drops into the higher air pressures below. These winds are known as "Chinooks" or "Santa Anas" in the western United States. The higher temperatures have been blamed for forest fires and a host of malfunctions associated with over-use of air conditioners. The winds of the Chinooks often reach terribly destructive speeds.

Windspeed Characteristics

Winds at virtually all sites have some remarkably uniform characteristics. At any

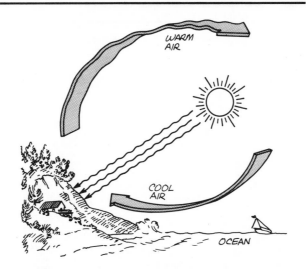
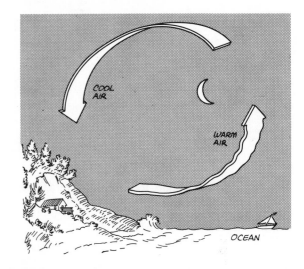

Sea-land winds occur because adjacent land and ocean masses have different rates of heating and cooling. On-shore breezes occur during the day and off-shore breezes at night.

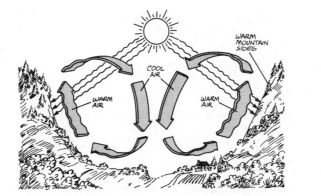
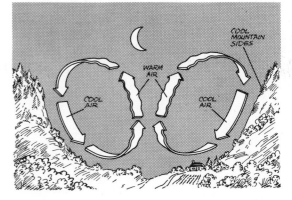

Mountain-valley winds. These gentle breezes occur because the air over high mountain slopes warms up faster by day and cools down more rapidly at night.

given site, there will be a length of time when there is absolutely no wind. For some other length of time, the wind will blow at an average speed, and for another length of

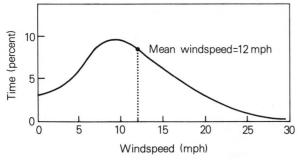

Windspeed distribution for a typical site with 12-mph average winds. This graph indicates the percentage of time, over a period of one year, that the wind blows at any given speed.

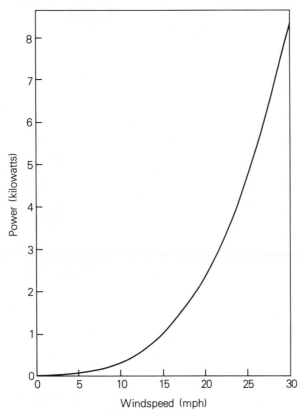

Available wind power as a function of windspeed. An ideal efficiency of 59.3 percent has been assumed for a rotor with frontal area of 100 square feet.

0 mph to its maximum, you could easily plot a graph like this one. The vertical axis can be expressed either as the total time or as the percentage of time the wind occurs at each speed. The shape of the curve will probably differ for your site, but the message will be similar.

Whenever many measurements of the windspeed are made, an average of those measurements should be calculated. Simply add up all of the measurements and divide that total by the number of measurements taken. At this hypothetical site, the *average windspeed* (or, equivalently, the *mean windspeed*) is about 12 mph—a few miles per hour greater than the most frequently occurring windspeed, about 9 mph for this site.

Why should you measure the windspeed in such detail? Why not just measure a simple average windspeed instead? The answer to these questions depends on how much reliance you expect to place on your wind system and how much time and money you have to spend. Let's look at windspeed characteristics more closely to see how the illustrated curve—or a similar curve for your site—relates to wind energy production.

Remember that the available wind power increases as the cube of the windspeed. We can use Equation 1 to calculate the maximum wind power that might be extracted at each of the speeds shown on the windspeed distribution curve. For example, the top right graph indicates the maximum wind power could be extracted by a rotor

time—perhaps only a few minutes over an entire year—it will blow at its maximum speed. The top left graph shows a typical *windspeed distribution* that will help you visualize these characteristics. If you measured the windspeed at your site for an entire year and then added all of the minutes the wind blew at each of the different speeds, from (assuming an efficiency of 59.3 percent) with an area of 100 square feet, when the windspeed varies from 0 to 30 mph.

This graph illustrates how dramatically wind power increases with increasing speed. At 10 mph only 300 watts can be extracted, but at 30 mph more than 8,000 watts, or 8 kilowatts, could theoretically be harnessed.

Those two graphs describe two major wind characteristics: statistical and power. The statistical characteristic is the amount of *time* you can expect each windspeed to occur, and the power characteristic is the amount of power you can expect to be available at each windspeed. Now remember that energy is calculated by multiplying power by time. A 100-watt light bulb left on for ten hours consumes 1,000 watt-hours, or 1 kilowatt-hour, of energy. The maximum energy available at your wind site is calculated in much the same way—by multiplying the power curve by the time curve. The result is an *energy distribution* curve, an example of which is presented here. This is the curve you really want to know.

At the hypothetical wind site, a windspeed of 10 mph occurs frequently. You might think that this speed will be of great importance in energy production. But the energy curve shows that it's much less important than windspeeds in the 15 to 20 mph range. That's where most of the energy is available. In fact, there is very little energy available below 8 mph or above 30 mph at this hypothetical site. The curves at any real site will probably differ from the ones in this

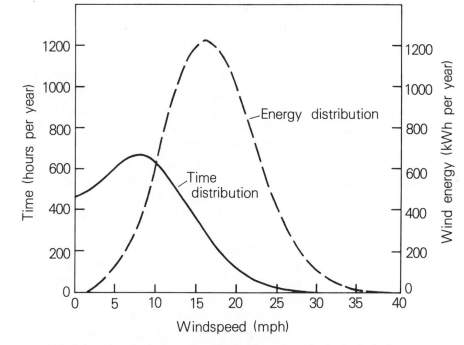

Windspeed and wind energy distributions for a typical wind site.

illustration, but the conclusions and the characteristics will be largely the same.

What other wind characteristics determine the shape of such curves? On a second-by-second basis, the wind is actually a succession of weak and strong pulses or gusts. On an hourly basis, the winds at a site have typical daily patterns—called the *diurnal variation* of the windspeed. At our hypothetical site, for example, the windspeed may be low in the morning, pick up in the afternoon, and reach its peak at about 8:00 p.m., as seen in the following graph. Two separate curves are shown for two differ-

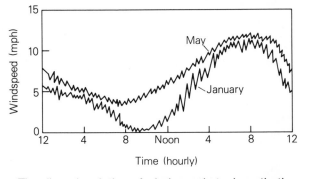

The diurnal variation of windspeed at a hypothetical site. May winds blow harder than January winds at this site.

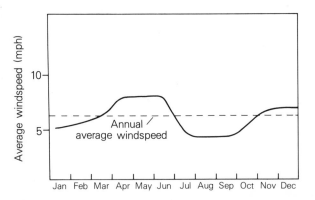

Monthly variation of windspeed at a hypothetical site.

ent months—January and May—to indicate that the diurnal variation may be different from month to month. Note that May is the windier month in this location, a fact that is also reflected in another plot of the month-to-month variation of the average windspeed.

These hypothetical curves indicate that windspeed varies on an hour-by-hour, day-by-day, and month-by-month basis. From a statistical viewpoint, the winds at any site can be described by a windspeed distribution curve, which tells you to expect a certain windspeed for a certain percentage of the time. It does not tell you when to expect that windspeed or how long it will persist. You can get a fair idea of when to expect certain windspeeds from a daily and monthly study of the winds at your site—if you choose to analyze them that closely.

By analyzing your site in terms of this daily and monthly variation, you will have an opportunity to compare energy needs with the wind energy available. At our hypothetical site, most of the wind energy is available in the evening. But suppose that the need for energy occurs in the morning. Some form of energy storage is required to retain the evening's production for the next morning's use. This might prove expensive. Suppose that the energy need occurs when the wind power is available. Little or no energy storage would be required, and the wind power system would cost less and perform better than if storage were needed. Hence, the wind power in this case is *worth* more. The value of a site's wind resource is directly

related to the energy available and to how well that resource coincides with energy needs.

How do you go about determining the windspeed distribution for your site? There are four basic options:

1. Ignore windspeed distribution entirely and base your design calculations on annual average windspeed and a correction factor.
2. Using the annual average windspeed at your site and a mathematical equation that describes the windspeed distribution fairly accurately, calculate the duration of each windspeed.
3. Actually measure and record the windspeed at your site for at least one year.
4. Measure and record the windspeed for a shorter period—perhaps three months—and try to establish a correlation with wind data from a nearby weather station or airport.

The *annual* windspeed distribution is, from the overall planning viewpoint, the most important factor to understand. Daily and monthly windspeed variations are, perhaps, the easiest to determine, but wind researchers increasingly favor an *assumed* annual windspeed distribution. You then make design calculations based on such an assumption, rather than actually measuring the windspeed for more than a year.

The *Rayleigh distribution* provides a reasonable description of windspeed characteristics in some locations. National Weather Service (NWS) wind data for several hundred locations have been compared with results from an assumed Rayleigh distribution. The comparisions have been promising, but there are a few problems of interpretation. Most of the NWS anemometers have been sited for monitoring airport winds, for making forest fire predictions and for a host of other uses unrelated to windpower production. Although the Rayleigh distribution doesn't work for all sites, it has been reputed to work with an error less than 10 percent. In the absence of better data, it can be used for reasonably good energy estimation.

The two graphs here show typical Rayleigh windspeed distribution curves for two sites with different annual average windspeeds. Notice that the energy content of these winds increases dramatically as the average speed increases from 10 mph to 14 mph. The vastly greater energy available at windier sites makes the required site analysis worth the effort. Energy distribution curves similar to these will be used later to design a wind machine that achieves optimum performance at windspeeds where the most energy occurs. In fact, peek windmill performance ought to occur at or near the same windspeed as the peak of the energy distribution curve.

Measuring the Windspeed

Measuring an actual windspeed distribution curve means taking many readings and filling many "bins" with this data. If you have a table covered with tea cups and you toss dried peas out onto that table you will be filling bins—in this case, tea cups. Throw enough peas out and a filling pattern will begin to take place that reflects the likelihood of that bin receiving a flying pea. The Rayleigh curves give a possible likelihood of any particular windspeed occurring.

If you look at a windspeed meter once each minute and add a "1" to the bin that corresponds to the windspeed you read, you are filling bins with minutes—the number of minutes the wind blows at each windspeed. Do this once each hour, and the bins contain hours. A simple daily reading will represent a "daily average" windspeed, and the number of readings in a bin will represent the number of days at a particular windspeed. A bin, then, contains the number of days, hours, or minutes that the windspeed happens to be measured at the value associated with that bin. More frequent readings will give a better representation of the actual windspeed distribution. One-minute readings are quite reasonable for electronic recording equipment, while hourly or daily readings are commonly taken by human meter readers at airport control towers, forest lookout stations, and other permanently staffed facilities.

Virtually all methods of calculating the annual average windspeed involve filling bins of one sort or another. If windspeed bins are filled electronically and accurately, the values in each bin can be used instead of the assumed Rayleigh distribution be-

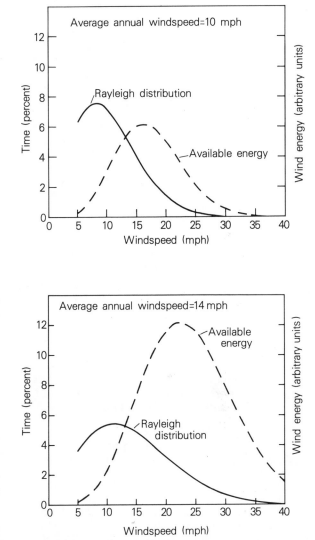

Rayleigh windspeed distributions for sites with 10 mph and 14 mph average winds. The available wind energy is far greater at the windier site.

The Rayleigh Distribution

With a year or more of windspeed data measured for a given site, you often need to find an equation that describes, within reasonable accuracy, the windspeed distribution. The most important parameter you should have by now is the mean windspeed. Within certain limits, a single parameter equation known as the Rayleigh distribution may be used to describe the windspeed distribution. Here the single parameter is mean windspeed. At windspeeds below 10 mph, the Rayleigh distribution has low reliability; it should not be used at all at sites with mean windspeeds below 8 mph.

The Rayleigh distribution takes the following form:

$$\text{Hours} = 8{,}760 \times \frac{\pi}{2} \times \frac{V}{\overline{V}^2} \times e^{-k},$$

$$(Eq.\ 3)$$

where V = windspeed
$\overline{V}$ = mean windspeed
$\pi = 3.1416$
$e = 2.718$
$k = \frac{\pi}{4} \times \left(\frac{V}{\overline{V}}\right)^2$.

This equation gives you the total number of hours per year you can expect the wind to blow at a windspeed V when the mean windspeed is $\overline{V}$ at that site.

A graph of this complex equation is presented at right, with percent time as the vertical scale. To get the result in hours per year, multiply by 8,760. Appendix 2.1 presents numerical values of the Rayleigh distribution for mean windspeeds ranging from 8 to 17 mph. You can read percent of time from this graph or consult the Appendix to get a more accurate value in hours per year. For example, for mean windspeed of 14 mph, you can expect wind to blow at 23 mph for about 2.2 percent of the time, or 194 hours per year.

The other graph here shows the agreement between a measured windspeed distribution and a calculated Rayleigh distribution for St. Ann's Head, England—with a mean windspeed of 16.2 mph. The Rayleigh distribution is slightly low at high windspeeds and high at low windspeeds (from 10 to 20 mph). As power is proportional to the cube of the windspeed, the higher speed end carries more weight in power calculations, but the greatly reduced duration of time at high windspeed reduces the overall energy impact. Rayleigh calculations are not recommended as a replacement for actually measuring your site's wind characteristics, but they can serve as a reasonable approximation when all you have is the annual average windspeed.

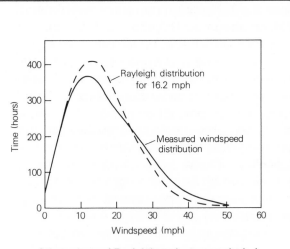

Comparison of Rayleigh and measured windspeed distributions for St. Ann's Head, England.

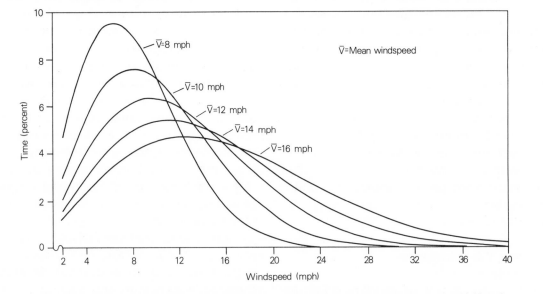

The Rayleigh distribution versus windspeed for sites with mean windspeeds between 8 and 16 mph.

cause they represent the entire windspeed distribution.

In another approach, you measure windspeed periodically during the course of a day, and average all the readings during that day to get the daily average windspeed. To arrive at an annual average windspeed, do this for a whole year, and average all daily readings. For maximum reliability, the readings should be taken at regularly scheduled intervals. As you might guess, this process becomes burdensome over an entire year.

A simpler alternative is to use a special device called a wind energy monitor. It adds up the total miles of wind that have passed the anemometer's sensor. Divide the total miles by the number of hours between readings on a daily, monthly, or annual basis, and you get daily, monthly, or annual average windspeed—simply and directly.

In addition, a wind energy monitor records the total wind energy available at a site. Each windspeed reading is converted directly into an energy value that is accumulated minute by minute. Such an approach eliminates the errors that might occur if you measured an average windspeed and later calculated the available wind energy using the Rayleigh distribution.

Wind Direction

Local winds are influenced by pressure and temperature differences across a few miles of land. These local atmospheric influences in combination with those of hills, trees, and other topographical features, cause wind to shift directions frequently—much as a flag waves about in the breeze. At any particular site, however, one general wind direction will prevail. This direction is called the *fetch area*. Sailboat skippers often use this term. Ask a long-term resident of the area in which you plan to site a wind system where the wind comes from. Chances are the answer will closely describe your fetch area. During a site analysis, you should become aware of structures, hills, or trees that might interfere with windflow. This can save a lot of work.

It's easy to visualize the impact of wind direction on site analysis. Suppose all of the data used to plot the windspeed distribution curve shown earlier were replotted as a three-dimensional graph to include wind direction. In the simplest version, this graph would show the relative amount of time that the wind blows at various speeds from the north, south, east and west. One could also multiply time by power as before to get the distribution of energy available from each direction. Both these distributions—time and energy content—are shown in the three-dimensional graphs here. They were generated from actual wind data gathered at the Palmdale airport in California.

The time curve shows that the wind blows mostly from the south and west at this site. But the energy distribution curve shows that the greatest amount of energy results from

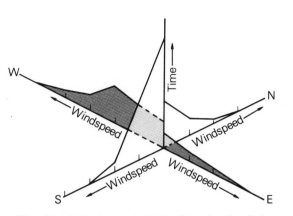

Directional windspeed distributions for the Palmdale, California, airport. Most of the time, the wind here blows from the south and west.

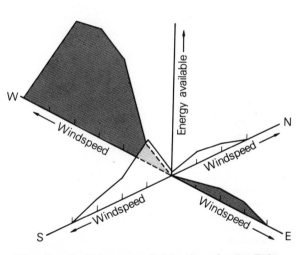

Directional wind energy distributions for the Palmdale airport. A large fraction of the wind energy comes from the west—the direction of highest windspeed.

westerly winds. For most wind sites, one direction will dominate like this. Such a three-dimensional visualization makes site analysis much easier.

How do you collect the data needed to construct such a graph? Make bins as before, but duplicate the bins for each of four, eight, or sixteen directions of the compass. The more directions, the more detailed your graph can be, but eight is usually sufficient. How is this done with site analysis instruments? An anemometer measures the windspeed and selects, on a time basis of once per minute, a windspeed bin. A wind vane, which senses wind direction, determines which of eight bins of the same windspeed will be incremented with one count. For example, suppose the windspeed measures 10 mph. There are eight possible bins marked 10 mph, one for each of eight directions. The wind vane decides which of those eight bins gets the count. This process is repeated each minute for a year; hence, 525,600 counts will be scattered among the bins. That's more than enough to construct a graph similar to those shown for Palmdale airport.

Wind Shear

The windspeed at a site increases dramatically with height. The extent to which windspeed increases with height is governed by a phenomenon called *wind shear,* a term derived from the shearing or sliding effect of fast-moving air molecules slipping over the slower ones. Friction between faster and slower air leads to heating, lower windspeed, and much less wind energy available near the ground.

The region of sheared air between unretarded air flow and the ground surface is known as the *boundary layer*. It has a definable and often predictable thickness. Accuracy of prediction depends on your ability to estimate surface friction factors or measure the windspeed at several heights simultaneously. Even wind flowing over a smooth surface will develop a boundary layer. The further the wind travels, the thicker the boundary layer. Minimum thickness occurs over a large, calm lake, or an ocean that

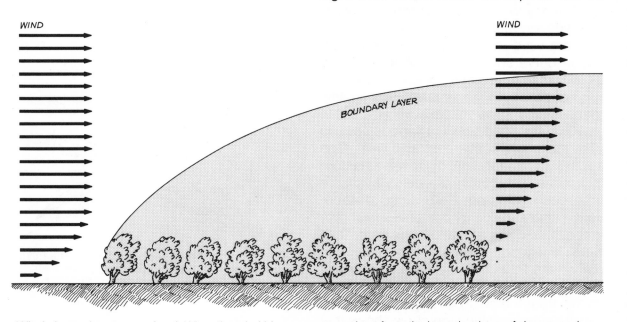

Wind shear above an orchard. When the wind blows over a rough surface, the boundary layer of slower-moving air thickens above it.

isn't subject to winds lapping waves and increasing surface roughness.

A more typical example of boundary-layer buildup and the effect on windspeed profiles is shown in the diagram on page 56: the wind approaches an orchard with a windspeed profile illustrated on the left. The trees extract some energy from the wind, and the profile on the right represents the wind leaving the orchard. A wind machine installed deep within the resulting boundary layer, say near tree-top level, would have much less wind energy available to it.

Windspeed profiles for three representative types of terrain are shown in the diagram on this page. The numbers along the curves represent percentages of maximum unrestricted windspeed occurring at each altitude. Over urban areas, the boundary layer is often more than a quarter-mile thick. But over level ground or open water, the wind reaches its maximum speed at less than 1,000 feet.

These percentages can help you estimate the windspeed to expect at one height if your anemometer is mounted at another. Suppose, using the "suburbs" curve, that your anemometer is mounted at the same height as the 60 percent mark (about 200 feet in the air) but you want to know what to expect at the 50 percent mark (about 100 feet up). The annual average windspeed measured by the anemometer will be 60 percent of the unrestricted annual average, and the windspeed at the lower height will be another 10 percent lower. To get the windspeed at 100-foot level, simply multiply the anemometer reading by the ratio of these two percentages. For example, if the annual average, as measured by the anemometer, is 10 mph, then the average at the lower height will be 10 times (50/60), or 8.3 mph. The height difference lowers the windspeed by 16 percent and the available wind energy by 43 percent. Thus, the wind energy available to your machine is very sensitive to the tower height.

The value of α (Greek "alpha") listed with each profile is the *surface friction coefficient* for that type of terrain. It represents an estimate of the actual surface friction near each site and is used in a formula to cal-

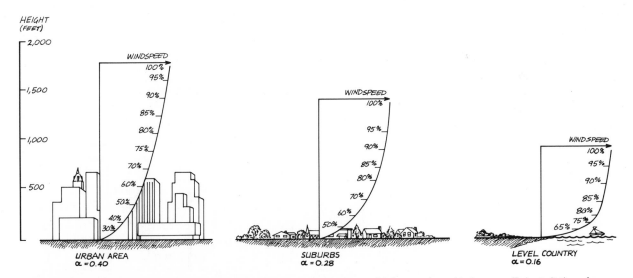

Windspeed profiles over different terrain. Rougher terrain has a higher surface friction coefficient; it therefore develops a thicker boundary layer above.

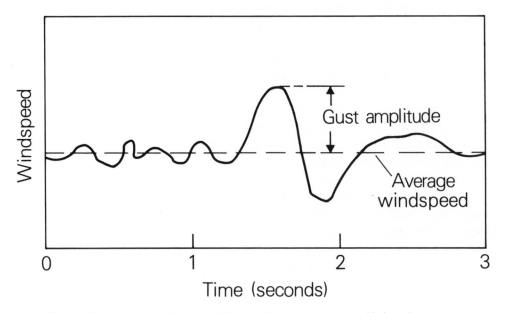

Graph of a typical wind gust. In this case, the departure speed is just the average windspeed.

NUMBER OF GUSTS PER YEAR TO TWICE THE DEPARTURE SPEED							
	Mean Windspeed						
Departure Speed	$(\alpha = 0.2)^+$				$(\alpha = 0.3)^+$		
	10	12	14	16	10	12	14	16
10	25	22	18	15	2269	2004	1701	1430
20	5	8	11	12	430	780	1022	1140
30	0	1	2	4	13	77	207	369
40	0	0	0	1	0	2	17	57
50	0	0	0	0	0	0	1	5

$^+\alpha$ = Surface Friction Coefficient

culate the effects of wind shear. This formula is presented in Appendix 2.2, along with more quantitative information about windspeed profiles.

Turbulence

An understanding of atmospheric turbulence is important for the structural design of a safe wind machine. Wind machines average the short-term pulses associated with gusts, so power output appears smooth, even though the actual windspeed is not. A very strong wind gust may destroy the machine.

What does a wind gust, or short-term turbulent variation, look like? A typical gust is shown in the accompanying graph. The increase in windspeed causes a theoretical increase in wind power, but usually this increase lasts for less than a second. A typical windmill cannot respond that quickly, and such short gusts have little effect on the power output.

For structural design, it's important to have some means of predicting the number of gusts you can expect at your site that have large amplitude. Usually it's enough to do this for several annual average windspeeds and for two general classes of surface friction. The table here gives the estimated frequency of gusts with twice the *departure speed* (the windspeed just prior to the gust).

Suppose your site has a fairly rough

terrain with α near 0.3, and the annual average windspeed at the site is about 16 mph. You can expect 1,430 gusts during the year that double the 10 mph departure speed to 20 mph. Notice also that you can expect up to five departures from 50 mph, that is, five gusts to 100 mph per year at this site. By comparison, at a site with a lower surface roughness, say $\alpha = 0.2$, and the same 16 mph average, you can expect only one departure from 40 mph to 80 mph, and none to 100 mph. Rougher surfaces induce gustier winds. In designing your machine, you can use the peak windspeeds from calculations like these in lieu of other data derived from long-term measurements at the site. Such peak windspeed information is essential for the structural design of the windmill blades and tower.

Site Survey

In a site survey, you head for the actual site selected, armed with a compass, note pad, tape measure and camera, and various anemometers. By way of comparison, in wind prospecting you start out equipped with all the same devices but do not necessarily know where the site is located. A site survey should provide you the data you need to plan a system for that site. Not too many years will pass before wind prospectors, armed with general wind maps and maps of existing electric power lines, will comb the windy areas of this country look-

ing for hot areas to "wildcat" wind-rights leases. These New-Age prospectors will instrument sites, get options on them, and sell their options to utility companies looking for good windy sites. To justify a sale, wind prospecting will require great care and accurate data-logging equipment.

Site surveying is much less rigorous than prospecting. The predictive tools, such as the Rayleigh distribution and gust table, allow you to perform simple surveys. But it's difficult to identify those areas where these tools are dependable. Tests show that a possible error of 10 percent or less is made by using simple statistical tools. But if these tools don't describe actual wind characteristics at your site, you can be off by as much as a factor of two on your energy estimate. For example, the Rayleigh distribution doesn't work well in regions with low average windspeeds—10 mph or less. These statistical tools are just no substitute for a good site survey using accurate, reliable instruments, although they do provide you with fair first-order estimates of the wind behavior at your site.

What makes a good wind site survey? It considers two things: the wind resource and the wind machine. In the first category you'll want to know:
- Annual average windspeed
- Windspeed distribution
- Wind direction
- Wind shear
- Surface roughness
- Site altitude.

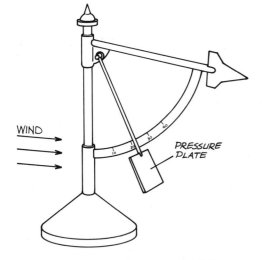

A pressure-plate anemometer. Wind pressure forces the plate to swing up along a graduated scale, providing a rough measure of the windspeed.

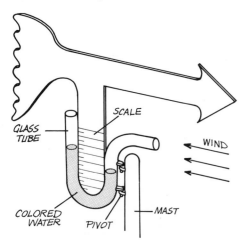

In an early pressure-tube anemometer, wind pressure induces a height difference between the fluid levels in a U-shaped tube.

Some site characteristics fall in both categories:

- Gale or tornado expectation
- Ice, sleet, hail, snow, and freezing rain
- Blowing dust
- Blowing heavy objects.

Other nonwind factors that can affect wind machine design are:

- Migratory birds
- Television interference
- Soil conditions
- Seismic stability
- Local social, legal, and environmental restrictions.

How do you conduct a site survey? What instruments are used? Site-survey questions concerning the wind resource are covered in the rest of this chapter. Siting factors unrelated to wind resources are discussed later.

Anemometers and Recorders

The simplest techniques for measuring include holding a wet finger up in the wind or tossing a fistful of fine sand above your head. Though these methods are not very accurate, one can hardly imagine a full-fledged site survey without them. A better indication of low-level air flow can be gained by using a child's bubble toy to disperse soap bubbles into the wind and watching them disappear. Streamers of yarn will substitute for bubbles. You are trying to obtain a three-dimensional image of the local wind.

More sophisticated instruments for measuring windspeed at a site fall into three main classes. These are:

- Pressure-plate anemometers
- Pressure-tube anemometers
- Rotation anemometers.

Pressure-plate anemometers seem to have been introduced around 1450 AD. Wind force on a plate swings it up against its weight. More wind, more sway. This same technique of measuring windspeed was used as an airspeed indicator on early barnstorming airplanes.

Pressure-tube anemometers were introduced around 1722. The diagram here shows an early "manometer tube," or pressure-tube anemometer. Wind blowing against the aimed tube (called a Pitot tube) exerts pressure against the fluid in the U-shaped tube. The height difference due to the fluid displacement under this pressure is read from the scale, and a chart is used to convert this difference to windspeed. At least one low-cost pressure-tube anemometer, the Dwyer Wind Speed Indicator, is available today with the fluid displacement calibrated in windspeed.

Patterned after windmills, rotating anemometers were introduced in the eighteenth century. These anemometers looked much like propellers with tail vanes to keep them aimed into the wind. In about 1846, the cup-type anemometer was developed. Four hemispherical cups were attached to radial arms, allowing the cups to spin about a vertical shaft. The cup-type, as well as the propeller-type, anemometers are read by measuring their instantaneous revolutions per minute (rpm) or by counting the total number of revolutions. By measuring the rpm, a direct read-out of windspeed is obtained. By counting total revolutions over a time period, typically one minute, you obtain an average windspeed over that time period.

Measuring the wind resource is only half of the job. Recording the wind data for future analysis is the other half. A low-technology approach to recording windspeed and direction, illustrated at far left, was first used around 1837. A flexible hose dispenses a fine stream of sand from a supported reservoir. The wind blows this hose away from the center of the ring. The relative sizes of the resulting piles of sand indicate both the magnitude and direction of the wind. Of course, this crude apparatus can only give qualitative estimates of average windspeed and direction.

Weather bureau measurements have been made on a read-once-per-hour, once-every-three-hours, or other similar basis. These readings are taken by a person reading various instruments and recording the values in a log. If reading is done on a once-per-hour basis, filling many bins with a great amount of data is an easy task—as long as a site is staffed full-time, as in an airport control tower.

If no full-time staff is available, such instruments as a strip-chart recorder will be required. This device can produce endless miles of paper marked by an ink line or

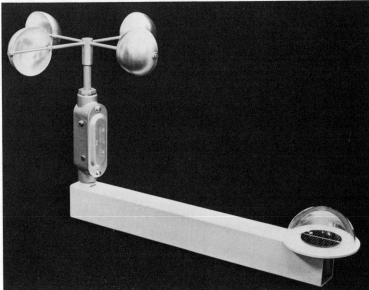

Cup-type anemometer with a solar pyranometer. Professional equipment like this must be used for detailed wind-energy measurements.

millions of tiny dots—each representing the average speed over about a one-second time period. But analyzing all this paper to obtain the windspeed distribution, annual average windspeed, energy content curve and other design information is a tedious task.

Recently, solid state technology has led to the development of wind analyzers that fill bins electronically. An example is the Helion E-400 Energy Source Analyzer shown in the bottom photograph on page 62. Even more recently, microprocessor technology has allowed significant cost reduction and per-

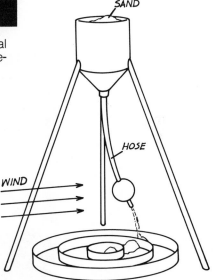

A "sandpile" anemometer, first used in the 1830's. This simple but crude method of recording windspeed provided only qualitative estimates.

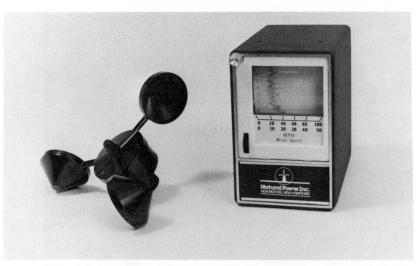

Strip-chart recorders are often used to log windspeed data when no staff is available to read monitors.

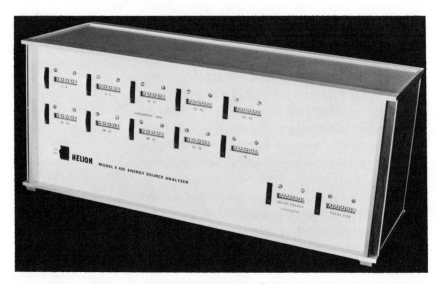

A modern electronic windspeed recorder. Solid-state electronics vastly reduces the labor involved in collecting and analyzing wind-energy data.

formance improvement in site recording instrumentation. Such data-loggers record dozens of channels of information and store this data on magnetic tape to be read by a computer that analyzes and summarizes the wind resource factors you need to design your wind system. The cost of these units has dropped to the point where they are economical for systems dealers and technicians. Using available low-cost computer kits, an electronically oriented person can build a data acquisition system suited to individual needs.

Regardless of the type of instrumentation you choose, certain instrument characteristics are of great importance and must be considered thoroughly. The linearity of the anemometer over the windspeed range you are studying is a critical factor. A nonlinear anemometer might read 15 mph when the wind is blowing at 15 mph, but read 31 mph at 30 mph for a 9 percent error in wind energy. Such inaccuracies must be understood or avoided. In steady winds, a good anemometer should be accurate to 1.0 percent of the value measured.

Most heavy-duty anemometers overestimate windspeed in gusty conditions. Anemometers that overestimate only slightly are made of balsa wood or light foam plastic and have little "coasting" or overshoot inertia. However, even a good metal or plastic anemometer will overshoot. General design refinements in rotating, cup-type anemometers have reduced the overestimation problem to a minor nuisance.

Any recording or measuring device yields the average windspeed over a time interval called its averaging period. For a strip-chart recorder drawing an ink line, the averaging period is determined by the quickness of the ink pen—perhaps less than a second. The dot-marker strip-chart recorders have an averaging period equal to the frequency at which dots are made—typically two seconds. An averaging period for a solid-state wind analyzer like the Helion E-400 is one minute. If the airport tower operator reads his anemometer once every hour, the averaging period is one hour. The shorter the averaging period, the denser the data and the better the actual description of the wind. For general siting work, a one minute averaging period is more than adequate. Data taken from equipment with short periods are usually averaged out over an interval of 15 minutes or more.

Other site instrumentation characteristics that should be considered include portability, remote battery life, immunity to extreme weather and lightning, and survivability from attack by wayward hunters. Historically, the last item is the most important!

Site Analysis

Anemometer in hand, you march out to the field where you expect to plant your wind machine. You will be asking yourself all sorts of questions regarding trees, buildings, turbulence, and such. Over the years,

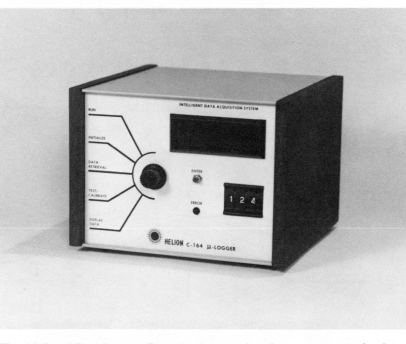

The Helion Micro-logger. Recent advances in microprocessor technology permit one to log dozens of channels of information for later computer analysis.

rules of thumb have evolved to help guide your first selection of a wind site. If possible you should select a site at least 20 feet above the tree or building height, and about 300 feet from the nearest obstructions.

Further rules of thumb depend on the wind fetch area. The long-time local residents are your best source of information about which way the wind blows. In some cases, permanent damage to local vegetation will tell the same story. You will hear stories like: "The rain winds always come from the south," or "Clearing winds come

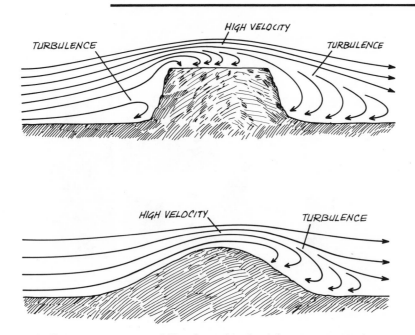

Air flow patterns over hills. Smoother land forms generate less turbulence.

through that pass over there." But what about other winds? "Oh, they don't amount to much." Local airport data may tell about wind directions. In fact, the main runway usually faces into the prevailing winds. Because of the influence of global circulation and large-scale winds, the wind fetch area can be predicted fairly well. Local breezes might travel some other path, but usually they don't play a major role in energy production. The rules of thumb mentioned should be used in siting relative to wind fetch area or according to prevailing winds. You will probably have to compromise on winds from other directions, but they will probably be of little value anyway.

A more detailed site analysis usually begins in the comfort of an office or home. Maps and weather data allow you to begin to examine your site—a circle or x on the map—and its wind fetch terrain. Available data may be from a source (an airport or weather station, etc.) exposed to the same winds and directly usable in evaluating your site. If so, the next phase of site analysis will verify it.

An anemometer should be installed at the site and data recorded over one to three months. These data should be compared with data from the same time period taken at the source near your site. If that source is much more than 50 miles away, correlation of data may not be possible. If a correlation does exist (for example, your windspeed is always 10 percent higher than theirs), you can use their long-term data directly, with your own correction factor applied to it.

In the absence of correlated long-term data, you must record your own. Six months is the minimum long-term measurement, and a year is advisable. The goal is to obtain acceptable values for the annual average windspeed, windspeed distribution, site roughness estimates, wind direction patterns, temperature trends, gustiness and the rest. Appendix 2.3 presents tables that summarize much long-term data and tell you how to obtain what information you need. Also, there are maps showing thunderstorm patterns so you can evaluate the site hazards they present.

Actual flow characteristics at your site are very complex. A qualitative assessment is possible using the accompanying illustrations of turbulent airflow. The purpose of the site analysis is to quantify wind flows in a general form usable for wind system planning and design. The rest of the book draws on this information as the design process unfolds.

The final result of a site analysis is information for windmill design (maximum and average windspeeds, wind shear and turbulence) and data for performance prediction (windspeed distribution, mean power and total energy measurements). Also, be sure to consider site factors related to installation safety and environmental effects (visual acceptability, migratory birds, television interference and the like). Overlook any factor and you risk an unsuccessful installation.

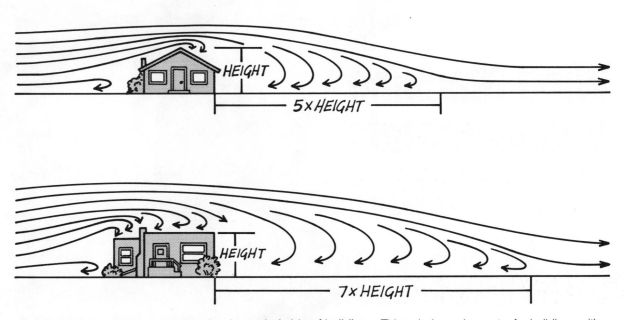

Substantial turbulence occurs on the downwind side of buildings. This turbulence is greater for buildings with sharp edges.

4
Wind Machine Fundamentals

Performance Characteristics of Common Rotors

A wind machine is any device that converts wind energy into other, useful energy forms. To remove kinetic energy from the air, its mass must be removed (I've not figured out how, but I'm sure it's illegal) or its speed reduced. Many things can reduce wind-speed and extract energy. Trees, for example, are better than solid fences because trees flex and dissipate wind energy within the trunk and branches. People have harnessed wind-driven tree motion to power water pumps by means of ropes, pulleys and springs.

Solid fences only create an obstacle around which air must pass, thereby losing only a small amount of energy to friction. Crash a car into a solid fence and you will convert *all* of its kinetic energy into heat energy and broken bones. Crash a bunch of air molecules into a fence and they pile up in front to form a ramp that allows the rest of the air to pass the fence virtually undisturbed. The best you can hope to do is slow the air down. That is the basis of windmill design: to create a machine that slows the wind and does something useful besides.

Two different types of wind machines have evolved that operate by slowing air down. The first type uses *drag* forces—much as the tree does. The second is a *lift*-type rotor that uses forces of aerodynamic lift. A familiar configuration for a drag-type wind machine is shown here. In this simple machine, kinetic energy in the wind is converted into mechanical energy in a vertical rotating shaft. One vane is pushed along by the wind

while the opposite vane moves against the wind around a circular path. The drag force on the latter vane must be overcome by the force on the first vane. Any extra force available is wasted unless a load is placed on the rotating shaft.

Suppose that a small electric generator is now driven by the power shaft. This generator will "load" the shaft, and the vanes will turn more slowly than an unloaded rotor under the same conditions. The downwind travelling—or power producing—vane will not be moving quite as fast as the wind. Thus, the wind will push harder on this vane.

If the shaft is held tightly and prevented from turning, no energy will be extracted from the wind, because the moving air will simply flow around the device and surrender only a small amount of its energy as heat. If the shaft is completely free, with no load impeding rotation, the machine will extract only the amount of energy required to push its vanes through the air—a small amount compared to that available. The vanes will spin very fast, and the machine will do very little useful work.

Lift-type machines use aerodynamic forces generated by wind flowing over rotor surfaces shaped much like an airplane wing. Lift force is generated *perpendicular* to the wind while a small drag penalty results that is parallel to the wind. Fortunately, the lift force is usually 10 to 50 times as strong as drag on the airfoil. The ratio of lift force to drag force, called the lift-to-drag ratio L/D, is an important design parameter. How does

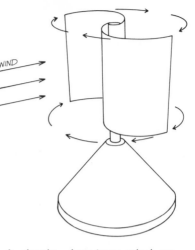

A simple, drag-type wind machine. Wind pressure on the high-drag, concave surface turns the rotor about its vertical axis.

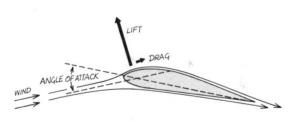

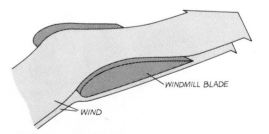

The flow of wind about a windmill blade. Lift forces act perpendicular to the local wind direction, while drag forces act parallel to it.

Drag-Type Machines

A drag-type wind machine harnesses the component of wind force perpendicular to the surfaces of its vanes. Such a machine might be a Savonius rotor or, even more simply, a flat board nailed to the end of a swinging arm. In this case, the drag force on the vane is given by the formula:

$$Drag\ Force = \tfrac{1}{2} \times \rho \times (V - u)^2 \times A_v \times C_D,$$

where

ρ = *the air density in slugs/ft*^{3},
V = *the windspeed in ft/sec,*
u = *the vane speed in ft/sec,*
A_v = *the area of the vane in ft*^{2},
C_D = *the drag coefficient of the vane.*

Generally, the drag coefficient of a vane has a value between zero and one.

If the rotor is at rest, the vane speed (u in the above equation) is zero, and maximum force occurs when the vane is perpendicular to the wind. If you multiply this maximum drag force by the radius to the center of rotation, you get the starting torque supplied by the vane. Of course, the net torque of the entire machine will be less because the wind is pushing against other vanes on the upwind side of the machine and retarding this rotation.

The power developed by a drag-type machine is just the drag force multiplied by the vane speed:

$$Power = \tfrac{1}{2} \times \rho \times (V - u)^2 \times u \times A_v \times C_D.$$

As the vane speed increases, the drag forces drop sharply (see graph), but the power extracted from the wind increases. When the vane speed equals one-third the free-stream windspeed V, maximum power extraction occurs. Of course, you still have to subtract the power wasted in driving other vanes

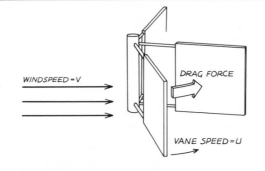

WINDSPEED = V

DRAG FORCE

VANE SPEED = u

upwind on the other side of the machine.

The drag coefficient C_D for a curved, two-vane Savonius rotor is about 1 for the concave, or torque, side and from 0.12 to 0.25 for the opposite, upwind-moving side. With these numbers you can easily calculate the difference in drag force between the two sides and estimate the net torque on the device. But be careful. Note that you should use V + u instead of V − u on the upwind vane. By a similar procedure, you can also estimate the net power developed by a Savonius.

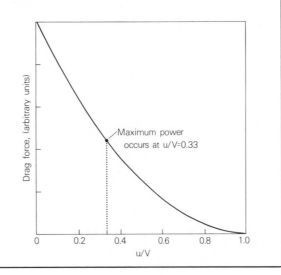

Drag force, (arbitrary units)

Maximum power occurs at u/V=0.33

0   0.2   0.4   0.6   0.8   1.0
u/V

lift produce the thrust which pushes the blade against its load? Note that the airfoil illustrated on page 67 is moving at an *angle of attack* off the relative wind. Lift is pointed slightly in the forward direction and, because the airfoil has a high lift-to-drag ratio, a net forward thrust results. This thrust tugs the blade along its rotary path.

Wind Machine Characteristics

All windmills have certain characteristics related to windspeed. At some low value of windspeed, usually from 6 to 12 mph, a windmill can begin to produce power. This is the *cut-in windspeed,* where the force of the wind on the vanes begins to overcome friction and the rotor accelerates enough for the generator or crankshaft to begin producing power. Above this speed, the windmill should generate power proportional to the windspeed cubed, according to Equation 1. At some higher speed, say 25–35 mph, wind loads on the rotor blades will be approaching the maximum strength of the machine, and the generator will be producing its maximum or *rated power.* A maximum useful windspeed, sometimes called the *rated windspeed,* will have been reached. It may also be the *governing windspeed,* at which some form of governor begins to hold power output constant, or even reduce power output at higher windspeeds. At some very high windspeed, say 60 to 100 mph, one might expect complete destruction of the

Forces on an Airfoil

All airfoils, even flat boards tilted into the wind and used as lifting surfaces, have predictable lift and drag characteristics. Lift is the force produced on the airfoil in a direction perpendicular to the "relative wind" approaching the airfoil. This relative wind is the wind that an observer sitting on the airfoil would face. The aerodynamic lift can be calculated from the formula:

$$Lift = \tfrac{1}{2} \times \rho \times V_r^2 \times A_b \times C_L ,$$

where

ρ = *the air density in slugs/ft³,*
V_r = *the speed of the relative wind approaching the airfoil, in ft/sec,*
A_b = *the surface area of the airfoil or blade, in ft,*
C_L = *the lift coefficient of the airfoil.*

The drag force on the airfoil occurs in a direction parallel to the relative wind; it acts to retard the forward motion of the airfoil. Its value is calculated by replacing the lift coefficient in the above equation by the airfoil drag coefficient, C_D.

To understand airfoils in more detail, you need to grasp a few other definitions. The "chord line" of an airfoil is a line extending from its leading edge to the trailing edge. The "angle of attack" is the angle between the chord line and the relative wind approaching the leading edge. The "pitching moment" is a measure of an airfoil's tendency to pitch its leading edge up or down in the face of the wind. It is important to the structural design of the blades and feathering mechanism. Certain airfoils are neutral; they have no pitching moment.

The graph presented here gives values of the lift and drag coefficients for a particular standard airfoil shape—the FX60-126. Similar curves are available for every airfoil tested.

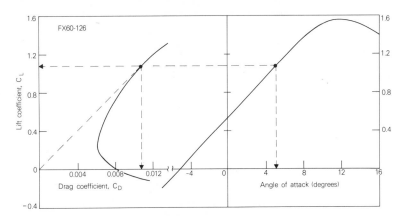

The curves shown here give the lift coefficient C_L versus angle of attack and include a "drag polar" that shows how the drag coefficient C_D varies with the lift coefficient. Note that the maximum lift occurs when the angle of attack is 12° and that the minimum drag occurs at $C_D = 0.006$, corresponding to a lift coefficient $C_L = 0.2$.

Example: *At an angle of attack equal to 4°, the FX60-126 airfoil has a lift coefficient $C_L = 0.96$. What is the lift force produced if the windspeed at the leading edge equals 40 mph and the blade area is 2 square feet?*

Solution: *First convert 40 mph to 58.8 ft/sec by multiplying by 1.47. Then, using the above equation for the lift force,*

$$Lift = 0.5 \times 0.00238 \times (58.8)^2 \times 2.0 \times 0.96$$
$$= 7.9 \text{ pounds .}$$

From the graph, $C_D = 0.0098$ when $C_L = 0.96$, so the drag force on the airfoil under the same conditions is:

$$Drag = 0.5 \times 0.00238 \times (58.8)^2 \times 2.0 \times 0.0098$$
$$= 0.081 \text{ pounds .}$$

By taking the ratio of the lift force to the drag force, you can calculate the lift-to-drag ratio, L/D:

$$L/D = \frac{7.9}{0.081} = 98 .$$

Of course, this is the same result you would obtain if you just took the ratio of the lift coefficient to the drag coefficient.

The best airfoil performance occurs at an angle of attack where the lift-to-drag ratio is a maximum. There you get maximum lift for minimum drag, but not necessarily the absolute maximum possible lift. On the FX60-126 airfoil, note that minimum drag occurs when $C_L = 0.2$—a low value compared to the maximum possible ($C_L = 1.6$). To find the angle of attack at which L/D is maximized, simply draw a line from the origin of the drag polar curve to the point where it just touches tangent to this curve. The point of tangency corresponds to maximum L/D for the airfoil. Draw a horizontal line from the point of tangency right to where it intersects the lift coefficient curve, and you get $C_L = 1.08$ in this example. As the drag coefficient here is $C_D = 0.0108$, the lift-to-drag ratio has a maximum value of 100. Note also that the angle of attack for maximum L/D is 5.2°. Setting the blade edge at this angle of attack to the relative wind will allow the airfoil to fly at its optimum performance.

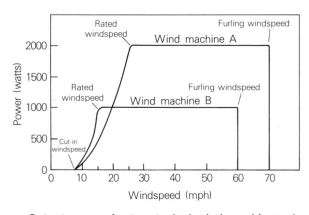

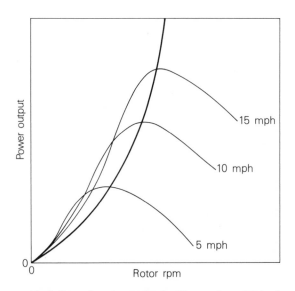

Output power for two typical wind machines. A rotor produces its maximum power at windspeeds between the rated and furling windspeeds.

Variation of power output with rpm for a typical rotor. At each windspeed, there is a point of optimum performance (heavy line).

machine if it were permitted to continue generating power. Wind loads on the blades or structural members will have surpassed their material strength, and catastrophe is the only possible result. The machine is usually shut down entirely before that, at a speed called the *furling windspeed.*

The characteristics for two hypothetical wind machines are illustrated in the accompanying graph. Machine A is a 2-kW machine with a rated windspeed of 25 mph, and machine B is a 1-kW machine rated at 15 mph. Machine B has a smaller diameter than machine A and is perhaps more fragile—its recommended furling speed is 60 mph, as compared with 70 mph for machine A.

These characteristics are very important. You have complete control of most of them during the design process. You first select the rated windspeed and power output. By designing for a given structural strength, you can calculate when furling *must* occur. You really cannot calculate the exact cut-in speed. It is as much determined by blade aerodynamics—which you can calculate— as it is by the thickness of oil in the transmission, bearing friction, and the phase of the moon.

Let's use the drag-type Savonius rotor to illustrate how rotors can be overloaded, underloaded, or loaded to their optimum power output by a generator or other load. The generator that loads the power shaft might draw enough power to overload the shaft and slow the rotor rpm to the extent that most of the wind just piles up and flows

around the machine—causing efficiency and power output to drop. Or, the generator might not extract enough power, and the rotor will spin too fast—causing extra drag on the upwind vane, lower efficiency, and added power loss. Somewhere between overload and underload is the *optimum load.* This optimum load is the extracted power that you calculated in Equation 1. All you need is the windspeed and the size and efficiency of the machine. The first two are fairly straightforward, but the last one depends on a number of factors that are discussed in more detail in the box on page 72.

Suppose you want to study more closely how wind power and rotor loading are related. How would you represent the relationships of loading, windspeed, and windmill performance? The bottom graph illustrates these factors. For a hypothetical wind machine, rotor power output is plotted against rotor rpm for several windspeeds. For example, the curve for a windspeed of 5 mph shows how power output at optimum loading is much greater than for overload or underload conditions (which allow the rotor to underspeed or overspeed, respectively). For the 10 mph and 15 mph curves, the effect is the same but stronger. Connect the peaks of the power output curves and you get the optimum load power curve for that rotor.

What causes the shapes of the peaked curves? Each curve gets its shape from the response of the rotor to loading and to wind gusts. Some rotors respond well, with a somewhat flat-topped curve. Such rotors are insensitive to non-optimum loadings or wind gusts. A large change in rpm means only a small change in power. Other machines might be so sensitive that slight overloading "stalls" the rotor—it quits turning altogether. You would expect the performance curve for such a rotor to have a sharply peaked shape. A small change in rpm can mean a large change in power output for constant windspeed.

In our discussion of rotor performance, the term *tip-speed ratio* (TSR) will often be used instead of rotor rpm. The TSR is the speed of the rotor tip (as it races around its circular path) divided by windspeed. For any given windspeed, higher rpm means higher TSR. If the tip is travelling at 100 mph in a 20 mph wind, the TSR = 5. Typical values of the TSR range from about 1 for drag-type machines to between 5 and 15 for high-speed lift-type rotors. By using the tip-speed ratio we can ignore the rotor rpm and diameter, and consider rotor performance in a more generalized discussion.

Wind Machine Performance

The basic formula used in calculating wind machines. Notice that the American Farm multibladed machine and the Savonius rotor are both low-TSR machines, operating at a TSR close to 1. The high-speed two- and defined earlier as rotor power output divided by power available in the wind. The efficiency of a wind machine depends on its design,

on how carefully that design is built, and on whether the machine is optimally loaded. No matter how well-designed and built, if a windmill is overloaded or underloaded it loses efficiency. In a plot of efficiency versus tip-speed ratio for several wind machines, each curve shows a distinct peak corresponding to optimum loading. The response of the machine to overspeeding and underspeeding of the rotor is indicated by the dwindling efficiency on either side of the peak. The graph here shows how efficiency—also called the *power coefficient,* Cp—relates to the tip-speed ratio for several types of wind machines. Notice that the American farm multibladed machine and the Savonius rotor are both low-TSR machines, operating at a TSR close to 1. The high-speed two- and

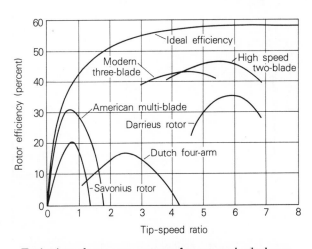

Typical performance curves for several wind machines. Rotor efficiency is the percent of available wind power extracted by the rotor.

Tip-Speed Ratio

The tip-speed ratio, or TSR, is a term used instead of rotor rpm to help compare different rotors. It is the ratio of the speed at which the blade tip (the furthest point from the center of rotation) is travelling to the free-stream windspeed:

$$\text{Tip-Speed Ratio TSR} = \frac{\text{Blade Tip-Speed}}{\text{Windspeed}}.$$

If you know the windspeed, the rotor diameter or radius and its operating rpm, you can calculate the tip-speed ratio, or speed ratio SR at any fixed radius between the center of rotation and the tip:

$$\text{Speed Ratio} = \frac{2\pi \times r \times N}{60 \times k \times V},$$

Or

$$SR = 0.105 \times \frac{r \times N}{k \times V},$$

where

   N = rotor rpm,
   r = radius at which SR is being calculated,
   V = free-stream windspeed, in ft/sec,
   k = a constant to adjust V:
      k = 1.47 if V is measured in mph,
      k = 1.00 if V is measured in ft/sec.

To calculate the tip-speed ratio with this equation, just use r = R (radius of blade) = ½ × D (rotor diameter).

   Example: *A rotor turns at 300 rpm in a 15 mph wind. If its diameter is 12 feet at the tip, calculate the TSR.*
   Solution:

$$r = R = \frac{12}{2} = 6\ ft;$$

$$TSR = 0.105 \times \frac{6 \times 300}{1.47 \times 15} = 8.6$$

The blade tip travels 8.6 times as fast as the wind.

Maximum Rotor Efficiency

The analysis of maximum possible efficiency for lift-type rotors was originally done by Betz in 1927. Here, the rotor extracts power from the airstream by slowing down the free-stream windspeed V to a lesser speed V_2 far downstream of the rotor blades. The power extracted is just the difference in wind energy upstream and downstream of the rotor, or

$$Power = \tfrac{1}{2} \times M \times (V^2 - V_2^2) \, ,$$

where M is the mass of air that flows through the rotor per second. If V_2 equals zero in the above equation, you might expect that power would be maximized. But no air would flow through the rotor in this case, and the power is zero. The mass flow through the rotor is just the air density times the rotor area times the average wind velocity at the rotor, or:

$$M = \rho \times A \times \frac{V + V_2}{2} \, .$$

Substituting this formula into the power equation yields:

$$Power = \tfrac{1}{4} \times \rho \times A \times (V + V_2) \times (V^2 - V_2^2) \, .$$

A graph of the relative power generated versus the ratio of V_2 to V is presented here. Note that maximum power occurs when V_2 equals one-third of V. Under such conditions,

$$Maximum\ Power = \tfrac{1}{2} \times \rho \times A \times V^3 \times \frac{16}{27} \, .$$

Thus, maximum possible (theoretical) efficiency of a lift-type rotor is 16/27, or 59.3 percent. In reality, swirl in the downwind airstream and other inefficiencies limit the practical efficiency even more.

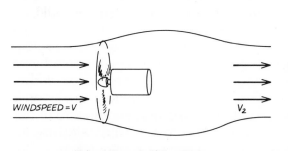

WINDSPEED = V V_2

Airflow through lift-type rotor.

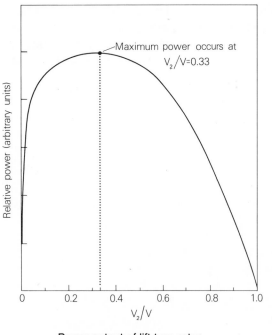

Maximum power occurs at $V_2/V = 0.33$

Relative power (arbitrary units)

0   0.2   0.4   0.6   0.8   1.0

V_2/V

Power output of lift-type rotor.

three-blade machines operate at high TSR, from 4 to 6, and higher efficiencies.

In Chapter 3, you saw that wind is actually a series of individual gusts. With this in mind, suppose that a Dutch four-arm windmill is spinning in a continuous 10-mph wind and a generator is loading the rotor to its optimum power output. The tip-speed ratio equals 2.5 in this steady wind; that is, the tip of a vane is moving at 2.5 × 10 mph, or 25 mph. Now add the gusts. Suppose the first gust passes the rotor and doubles the windspeed to 20 mph. For a brief moment, the new tip-speed ratio is 25 ÷ 20, or 1.25. At this same instant, rotor efficiency drops to about half its original peak value, but the doubling of windspeed means that eight times as much windpower is available to the rotor. The actual power output only quadruples (½ × 8 = 4).

Because the rotor doesn't speed up instantly, it is actually averaging the effects of gust-induced variations in the tip-speed ratio. Over a long time period, a rotor whose efficiency curve drops off steeply on either side of the peak is less apt to convert as much wind energy as one whose efficiency curve is relatively flat. A rotor with a flat efficiency curve is insensitive to gusts. An important point to consider with efficiency curves is the change in performance that can be expected from halving or doubling the tip-speed ratio. Both height *and* shape of the efficiency curves are important design considerations.

In Chapter 3, the energy content in the wind was shown to be a peaked curve. As much as possible, the peak operating efficiency of a wind machine should coincide with the peak of the wind energy distribution. It's not always possible to have them coincide exactly, because a different wind machine cannot be designed for every individual site. But try to make the rotor efficiency curves look like, and peak at about the same windspeed as, your site's wind energy distribution curve. This visualization is the first step in selecting appropriate operating characteristics for your wind machine.

The accompanying diagram shows how wind energy distribution and rotor efficiency curves might coincide. The energy distribution was calculated from the windspeed distribution curve. Note that most of the wind energy is available at windspeed "A", while the rotor efficiency peaks at TSR "B", corresponding closely with windspeed A. As the TSR is a ratio of tip-speed to windspeed for fixed rotor diameter, this relationship determines the optimum rotor size for this hypothetical site.

So far I have discussed only the rotor and its efficiency. Overall efficiency of the wind machine is related to the actual performance characteristics of any component that can rob the wind machine of power. Often the bearings and transmission have losses that can be considered constant, but the rotor, generator, and other loads such as pumps have efficiencies that vary with windspeed, rpm, and TSR. These all combine to give the performance curve its final shape.

Types of Machines

A major factor used to classify the various types of wind machines is the method of rotor propulsion; the rotor is propelled either by drag forces or by aerodynamic lift. The first rotor discussed in this chapter uses direct impact of the wind against a vane to provide motive force. This machine depends on a difference in drag between the power-producing vane moving downwind and the opposite vane, moving upwind. The curved shape of the vane permits this difference in drag forces. But for power production, the vane tip-speed cannot be much faster than the windspeed. Otherwise, the vane would be moving away from the wind that is supposed to be pushing against it (not very likely). So drag-type wind machines operate best at a TSR close to 1.

Lift-type, or airfoil, rotors use the aerodynamic lifting forces caused by air flow over blades shaped like airfoils to turn the rotor. Smooth air flow over an airfoil produces lift that pulls the blade in the thrust direction. Simultaneously, a small drag force acts against this thrust. Drag is the penalty one must pay for hanging anything out in a breeze. Well-designed airfoils don't have anywhere near the drag of such unsophisticated shapes as flat boards. Lift-type rotors are not restricted by any limitations on the tip-speed ratio. In general, the higher the tip-speed ratio, the higher the rotor efficiency.

There are four generic types of wind machines discussed here: the Savonius

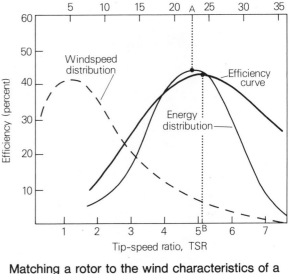

Matching a rotor to the wind characteristics of a site. The maximum efficiency of a well-matched rotor occurs at about the same windspeed as the peak in the wind energy distribution.

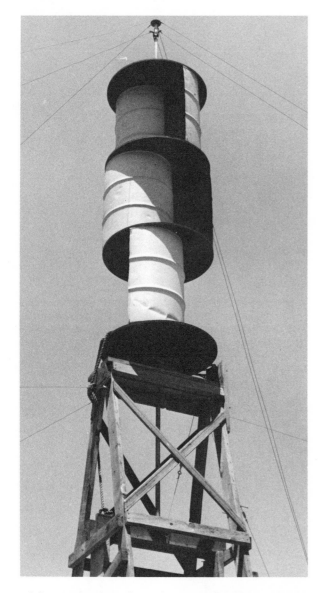

A low-technology Savonius rotor. Easily fabricated from surplus oil drums, this drag-type machine offers only limited power.

rotor, Darrieus rotor, multibladed farm windmills, and highspeed propellor-type rotors. These are the types most often encountered in design discussions and in the field.

Each of these types has evolved to serve specific needs or conditions. Savonius rotors and farm windmills are slow-turning with high starting torque—which suits them well to mechanical tasks such as lifting water. The Darrieus and propellor-type rotors spin much faster and have little or no starting torque at all. Their higher rpm make them well-suited for driving electric generators. The range of actual design types is vast, and even includes wind generators with no moving parts. Before you can make a thorough evaluation of your wind system's efficiency, you must select a design type that will form a basis for your analysis.

Savonius Rotor

The Savonius rotor, or S-rotor, looks something like an oil drum that has been sliced in half and separated sideways, as shown in the photo. It was officially invented by Sigurd J. Savonius of Finland in the early 1920's, although it was probably built by many other experimenters prior to that time. The rotor was originally developed to power specially designed sailing ships then being tested. The Savonius is a drag-type rotor. In addition to drag on the vanes producing rotary shaft power, that drag produces downwind forces (also called drag loads) on the tower.

A rotor that slows air down on one side while speeding it up on the other, as does the S-rotor, is subject to the Magnus Effect: lift is produced that causes the machine to move in a direction perpendicular to the wind. Spin on a baseball causes it to curve because of the Magnus Effect. An S-rotor can easily experience lift forces equal to two or three times the drag load placed on its supporting tower. Many owner-built S-rotors have toppled to the ground because their designers overlooked this phenomenon.

Recent theoretical studies have shown that the rotor efficiency of an S-rotor will most likely be less than 25 percent. If you add water-pump losses and other equipment inefficiencies in calculating overall efficiency, that's a maximum system efficiency of 15 percent for pumping water with a Savonius. Tip-speed ratios are about 0.8 to 1.0 at peak efficiency—as you would expect with a drag-type rotor.

The desirable features of the S-rotor are as follows:
- Easily manufactured by owner-builders
- High starting torque for starting under heavy load.

Undesirable features include:
- Difficult to control—other than a brake mechanism, controls to limit rpm in high winds are not readily devised
- Poor materials usage—presents a small frontal area for a fixed amount of construction materials

Often claimed, but not particularly important, is that the Savonius can convert energy from winds that rapidly shift direction. In most installations, the winds used for the major portion of energy production do not shift directions. The gusty, so-called energy winds are often stronger than windmills are normally designed to respond to. The big plus for an S-rotor is that it is easily built with readily available materials and can produce high torque while it is starting to spin. Thus, it is suited to a variety of direct mechanical uses such as pumping water, driving compressors or pond agitator vanes, and even powering washing machines, if that's your fancy. The number of vanes is not limited to two as shown here; three, four and more blades are common.

We saw earlier that a difference in drag force on the downwind-moving vanes to the upwind-moving vanes is needed to produce a net torque on the power shaft of a Savonius. By increasing this torque at the highest possible rpm, you can maximize the power output of this type of machine. There are two ways to accomplish this feat:

1. Maximize the difference in drag coefficients between upwind and downwind vanes, or,
2. Minimize the wind force against the upwind-moving vane.

Shapes that maximize this difference in drag coefficients have evolved mainly to the familiar Savonius rotor shape. Minor variations on this shape are possible with cones, wedges, or flat vanes that flop over edgewise as they advance into the wind. To minimize wind force against the upwind-moving vanes,

A three-tiered Savonius rotor designed to generate electricity.

Eggbeater-style Darrieus rotor being tested at Sandia Laboratories in Albuquerque, New Mexico. This high-performance machine uses extruded aluminum blades.

simply build a shield in front of them. With such a solution, you can use simple flat vanes rather than the more complex curved vanes.

Darrieus Rotor

Not long after Savonius patented his S-rotor, a French engineer named G.J.M. Darrieus invented another vertical-axis rotor. His patents anticipated virtually all of the major innovations being tried today with this type of windmill. Several Darrieus rotors are shown in the photographs on these pages. The two primary variations are the "eggbeater"—so named because of a distinct similarity in shape—and the straight-blade versions, sometimes called cyclo-turbines or cyclo-gyros by various developers of this design.

Both the Savonius and Darrieus rotors are *crosswind-axis* machines in which the power shafts are mounted either vertically or horizontally, perpendicular to the wind stream. But there is one important distinction: the Savonius is a drag-type device, while the Darrieus is a lift-type machine. The diagram on page 78 illustrates how lift forces on the blades act in a direction ahead of the blades, as all airfoils produce lift perpendicular to the airflow approaching the airfoil's leading edge. As the blade moves along its path, it is actually moving at a speed several times faster than the wind. Thus, even when the airfoil appears to be moving downwind, it is not. Lift is produced over almost the entire circular path. Contrast this case with the drag-type Savonius rotor, in which power-producing forces on the downwind-moving vane are fighting drag forces on the upwind blade vane. You can well imagine that the efficiencies of Darrieus rotors are greater.

Some theoretical studies indicate a 54 percent efficiency for the Darrieus rotor, not including losses in gears, generators, and elsewhere. Others think the Darrieus is actually capable of higher efficiencies than the theoretical maximum of 59.3 percent. There are good reasons for such claims; but neither of them has been proven correct, yet. In careful tests, the measured efficiencies ranged from 20 percent for the "egg beater" design, to greater than 50 percent for highly sophisticated straight-blade designs. This diversity of results suggests that the question is probably still open.

In any event, the efficiency curve for the Darrieus (see page 79) suggests the performance you might expect from a well-designed rotor. The steep slope on the low-rpm side of the curve indicates that this rotor is easily stalled when overloaded. Should the windspeed increase quickly while a fixed load is applied to the rotor, its tip-speed ratio falls rapidly. The rotor, which was operating at the peak of its performance curve, slips over to the steep underspeed side of the curve, even though more wind power is available to the rotor. Properly designed rotor and generator controls will prevent complete stalling of the rotor under this condition. Without such controls a rotor stall is almost guaranteed.

The desirable features of a Darrieus rotor are as follows:

- Possible ease of construction by owner-builders if lower performance is acceptable
- Low materials usage for high power output
- Adaptability to sail and other appropriate technologies
- Possible high wind-energy conversion efficiencies.

Undesirable features include:

- High-performance machines need complex controls to prevent rotor stall
- Difficult to start rotor.

Darrieus rotors are well adapted to driving electric generators or other high-speed loads. Because of the need to apply starting power to the rotor to accelerate it to high operating speeds, they are not well suited to lifting water directly or powering similar mechanical loads. In Chapter 2, however, a Darrieus used for pumping water in Bushland, Texas, was described. The rotor adds its power to the pump along with that of an electric motor. That same motor becomes an electric generator whenever the Darrieus is generating more power than is needed to pump water.

How a Darrieus Works

The Darrieus rotor works in an aerodynamic fashion similar to other lift-type rotors,

A straight-bladed Darrieus rotor. The pitch angle of the blades is changed automatically.

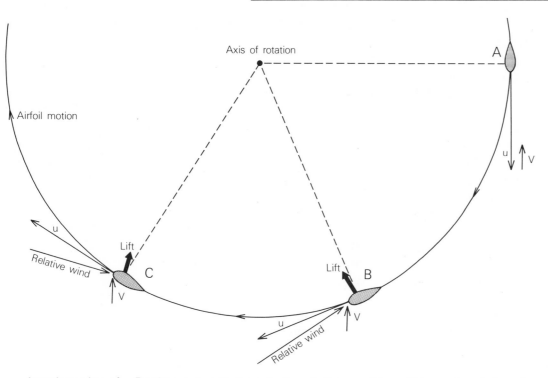

Aerodynamics of a Darrieus rotor. Under normal operating conditions, lift is produced along the entire carousel path, tugging the blades forward.

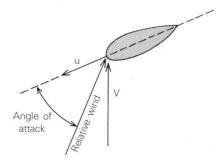

Low-speed aerodynamics of a Darrieus blade.

but because of the carousel path of the blades, its operation appears complex. It really isn't that difficult. Recall that airfoils generate lift perpendicular to a wind approaching the leading edge. In the Darrieus, that "relative" wind changes its angle from almost straight at the airfoil to nearly perpendicular to it. Thus, the amount of lift produced by the airfoil changes constantly as it sails around its path.

At position A in the diagram, blade motion and wind direction are parallel, although

pointed in opposite directions. No thrust occurs at this position. As the blade advances towards position B, however, the blade is at an increasingly steep angle to the wind. The lift force is directed ahead, along the direction of blade motion, and thrust is developed. Notice that blade speed is much greater than windspeed. So the tip-speed ratio is much higher than 1—maybe 5 or 6. A high tip-speed ratio is the key to successful operation of a Darrieus rotor. At a low TSR, the blade speed vector becomes shorter than the length of the windspeed vector, as illustrated in the next diagram. In this case, the angle of attack between the relative wind and the airfoil motion is too large, and stall can occur. Turbulent airflow, loss of lift, and high drag result in stall—obviously an undesirable condition. Compare the angle of attack that results from low TSR with the angle of attack in normal, high-TSR operation. At really low tip-speed ratios, stall is so prevalent that the rotor may require additional power from a starter motor to accelerate it up to operating speed.

Stall of the fixed-pitch Darrieus at low speeds results in an efficiency curve that looks like the one on page 79. At initial start-up, the rotor has a zero, or mildly positive efficiency. As rotation speeds up, stall effects rob the blades of power to the extent that the efficiency is actually negative. External power is usually required to accelerate the rotor through the stall region. Once beyond the stall region, acceleration is rapid up to the operational tip-speed ratio. Unless a gust suddenly drops the TSR back into this stall region, the Darrieus will continue to generate power unaided.

Under certain wind conditions, a Darrieus rotor can start without help. Peculiar, but not uncommon, wind gusts will accelerate a stationary rotor to operating speed. Often such a self-start occurs when the crew is off at lunch; nobody is around to see what happened. The result can be a thoroughly trashed rotor; if you don't expect the rotor to start, why hook up the load? Right? Absolutely wrong! Always expect a Darrieus rotor to self-start, even though the experts have told you it won't.

What are some of the alternatives for starting a Darrieus rotor? Electric starter motors are common. Such starter motors use a wind-sensing switch and a small electronic logic circuit to decide when it's appropriate to apply current to the motor. Another starting technique is to combine a Savonius rotor with a Darrieus. The Savonius has a high starting torque—enough to coax the Darrieus through its stall region. By making the Savonius just large enough for starting, it won't contribute to operational power.

An increasingly common starting method is that of *articulating* variable pitch blades that are hinged so that their pitch angle can change as they travel around the carousel path. The eggbeater is an unacceptable design for articulated blades; its curved blades cannot easily be hinged. The straight-bladed Darrieus can easily be hinged, and

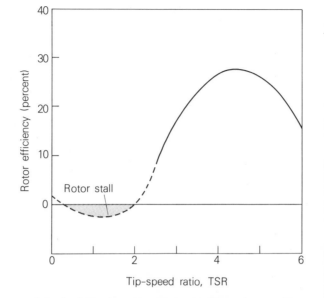

A typical Darrieus power curve. The rotor must be accelerated through a region of stall before it attains normal operating conditions.

it often is. To see how articulation works, start with the fixed-pitch blade diagram at blade position B. The blade is fixed exactly tangent to the circular path. At low TSR, this results in a high angle of attack.

Now, suppose that the blade pivots on its attached arms so that it points directly into the relative wind (i.e., its angle of attack equals zero degrees). Stall is eliminated, but so is lift. Optimum articulation lowers the blade angle to an angle of attack that produces maximum lift. But simple mechanical controls that articulate the blades often do not hold the blades precisely at optimum angles.

Small Savonius rotors along the axis help accelerate this straight-bladed Darrieus rotor through the stall region.

American Farm windmill in final stages of assembly.

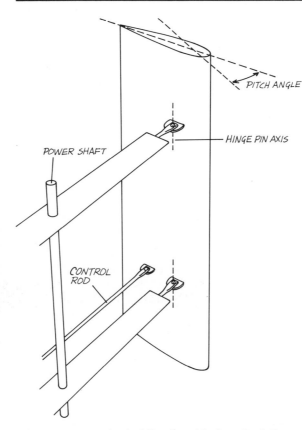

PITCH ANGLE

POWER SHAFT

HINGE PIN AXIS

CONTROL ROD

A common method of Darrieus blade articulation. Varying the blade pitch eliminates rotor stall.

The drawing above shows a typical structural configuration for articulating the blades of a Darrieus rotor. Each blade is supported from the power shaft by two arms. Each arm is attached to the blade with a hinge pin that allows the blade to pivot through the pitch angle illustrated. The blade is held at its pitch angle by a control link connected to any one of several control systems. The simplest control is a central wind vane that holds a cam in a position corresponding to the wind direction. The cam tells the control links to position the blades at a pitch angle approximating the optimum blade angle. Other methods of blade control usually involve electric or hydraulic servomechanisms driven by a small electronic circuit or computer. Whether fixed-bladed or articulated, the Darrieus rotor is very sensitive to its tip-speed ratio. Allowed to overspeed, the power coefficient drops until the lower power output equals the load. Overloaded or in a strong gust, the TSR drops, blade stall sets in on the fixed-blade machine, and the power coefficient drops severely on both types of Darrieus rotors.

Multiblade Farm Windmills

Multiblade farm windmills date back to at least the mid-1800's, when the earliest water-pumpers, built by Halliday, used flat wooden slats as blades. By the latter part of that century, the understanding of wind machine design was just entering an era of empirical and analytical aerodynamics. Eventually the flat wooden blades were replaced by curved sheet-metal blades, whose curvature improved their lift and also the energy conversion efficiency of the rotor.

The farm water-pumper evolved from a need for a high starting torque—it had to begin turning while lifting water at the same time. High torque at low rpm requires a large lifting surface area, so the total blade surface almost equals the windmill frontal area on these machines. An extra bonus of placing many blades close together is the so-called "cascade effect." Each blade aids the next by acting as a guide vane for the air flowing over its neighbor. The air flow at very low tip-speed ratios occurs at a nearly optimum angle for each blade to develop its maximum lift.

Another benefit of the cascade effect is that it tends to limit the maximum tip-speed ratio of the rotor. These rotors tend to operate at a TSR close to 1. At higher TSR, air flows into the rotor at angles far from optimum—thereby limiting the rpm. Some other means of shutting down the machine is still necessary during extreme high winds. Otherwise, the stresses exerted on the tower by the large surface area of this rotor become enormous. The rotor itself must be able to withstand such loads. A typical solution has been to tilt the rotor out of high winds.

The desirable features of the multiblade rotor are as follows:
- High starting torque
- Simple design and construction
- Simple control requirements
- Durability.

Its undesirable features include:
- Not readily adaptable to end uses requiring high rpm
- Exerts high rotor drag loads on the tower.

A farm water-pumper derives shaft power from the wind in a fashion remarkably similar

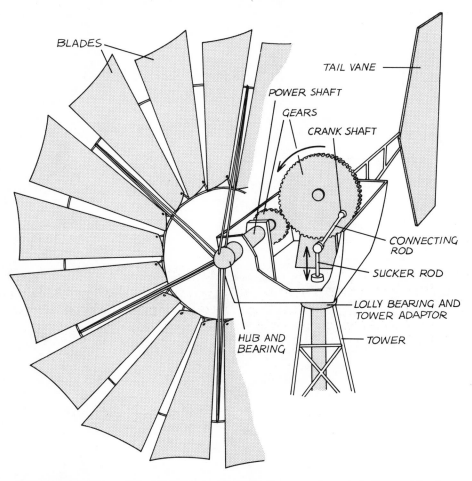

BLADES

TAIL VANE

POWER SHAFT

GEARS

CRANK SHAFT

CONNECTING ROD

SUCKER ROD

LOLLY BEARING AND TOWER ADAPTOR

TOWER

HUB AND BEARING

Components of an American Farm windmill. Gears and crankshaft convert rotary power into the up-down motion of the sucker rod.

to the high-speed aerodynamic rotor. But the farm water-pumper is not particularly sensitive to aerodynamic factors because of the cascade effect. With just one blade, you recall, the angle of attack can be steep

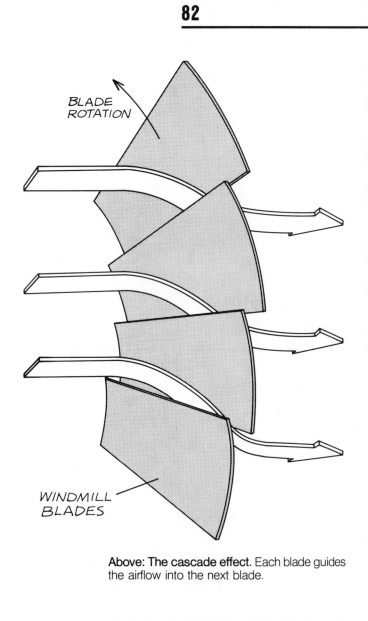

BLADE ROTATION

WINDMILL BLADES

Above: The cascade effect. Each blade guides the airflow into the next blade.

Right: Small wind machines like this Jacobs operate at moderate tip-speed ratios close to 5.

enough to stall the airfoil. But in the cascade system, each blade directs airflow into the next and reduces stall. The major penalty is a relatively high rotary speed, or *swirl,* added to the airstream downwind of the blades. Much less swirl occurs behind a high-speed rotor. Slightly lower efficiencies result because of swirl (as much as 10 percent loss in power), but the benefit of the higher starting torque is usually worth it. The maximum rotor efficiency attainable with a multiblade farm windmill is about 30 percent, but 15–20 percent is more common in practice.

High-Speed Rotors

High-speed, propeller-type rotors have a much lower solidity than farm water-pumpers and operate at much higher rpm. Two or three blades are common, and four a practical maximum. They begin operating at tip-speed ratios up to 5 and have been tested to about 20. Most factory-built wind generators operate in the 5–10 range, although a Hütter-Algaier 100-kW machine built in the 1950's operated at 16. These rotors are well suited for electrical generation because their high rpm lowers the gear ratio needed for driving a generator.

The design of high-speed rotors places much greater emphasis on blade aerodynamics than do the lower-rpm designs. Machines like the Jacobs, Kedco, and Wincharger operate at a moderate tip-speed ratio of about 5. At these ratios, good airfoil

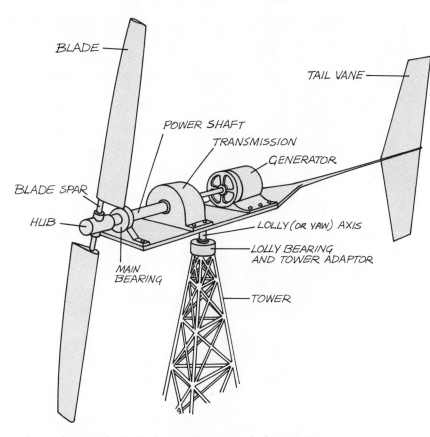

Components of a typical propeller-type wind generator.

design will improve the performance of a rotor by about 5 percent over one designed with less attention to aerodynamic details. Rotor efficiencies close to 45 percent are possible, and 40 percent is a common value in this range. At a tip-speed ratio of 8 or 10, close attention to aerodynamic details yields an efficiency of 45 percent, with a strong chance of lower transmission loss. The

The 6 kW Hütter-Algaier machine. The small, high-solidity rotor on this propellor-type machine uses crosswinds to face the main rotor directly into the wind.

Hütter machine attained overall efficiencies (including transmission and generator losses) of 40–50 percent at tip-speed ratios between 13 and 16.

The desirable features of high-speed, propellor-type rotors are as follows:
- Slender blades use less material for the same power output
- Higher rpm reduces transmission requirements
- Lower tower loads occur with slender blades
- Large diameters and high power levels are more easily attained.

Their undesirable features include:
- Lower starting torque
- Blades require very careful attention to aerodynamic design
- Possible flutter and vibration problems.

The high-speed rotor provides an excellent illustration of the aerodynamic relationship involved in coaxing power from the wind into a power shaft. The simplified diagram on the opposite page, which represents the cross section of a rotor blade turning in the wind, should help you visualize this process. Aerodynamic lift is produced at right angles to the relative wind that the airfoil "sees" at this particular cross section. This relative wind is the vector sum of the blade motion and the wind velocity at this position in the rotor disk. Blade lift tugs the blade along its rotary path, causing thrust. As long as the angle of attack is appropriate all along the entire blade, it will develop thrust at its maximum capacity. This thrust generates the shaft torque that spins a generator or drives a pump.

Some slowing of windspeed occurs in front of the rotor, so the windspeed at the rotor disk is less than the free-stream windspeed (i.e., the windspeed measured by a nearby, unobstructed anemometer). In an ideal rotor with maximum power extraction (59.3 percent efficiency), the windspeed far downstream of the rotor should be one-third the windspeed far upwind of the rotor. The windspeed at the rotor should be the average of this upwind, free-stream speed and the windspeed far downstream, or two-thirds of the free-stream windspeed. Thus, for maximum power the upwind slowing should be one-third of the free-steam windspeed. Ideally, this upwind slowing should be the same along the entire blade span, but on real rotors only a small portion of the span has this value.

Slipstream rotation or swirl—similar to that behind a multiblade farm windmill—occurs in the downstream airflow because the air is carried along with the blades as they travel their circular path. Although slipstream rotation can often be large, it is usually very small for high-speed rotors. Because of these induced flow factors, you cannot merely add vectors for wind velocity and blade motion to estimate relative wind by triangulation. You will be off slightly—with greater errors at low rpm than at high rpm. In Appendix 3.3 there are graphs that help you to calculate all important angles for blade design. The higher values of the blade speed

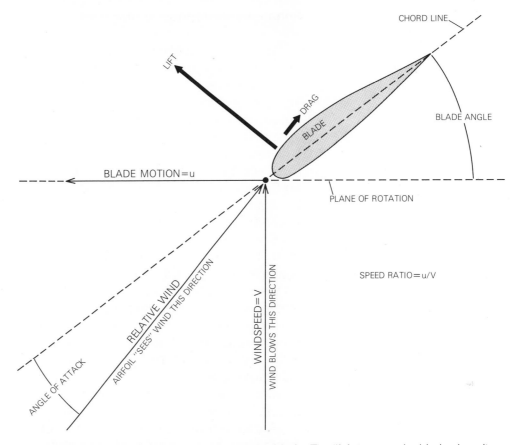

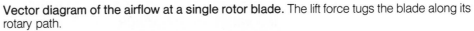

Vector diagram of the airflow at a single rotor blade. The lift force tugs the blade along its rotary path.

near the tip require some *twist* in a well-designed blade. Twist changes the blade angle as you move along the blade span from hub to tip. For rotors operating at tip-speed ratios from 3 to 8, twist is not so important. Above a TSR of 8 the twist, airfoil selection, and other design factors become increasingly important.

Enhanced-Performance Machines

You have probably seen photos or diagrams of strange-looking ducted rotors, vortex creators, and other odd machines. The general idea of these machines is to enhance the performance of the rotor by speeding up the wind flowing through it. One approach is to deflect more air into the rotor—perhaps with large vanes upwind of an S-rotor. A canyon or a pile of dirt in front of a Darrieus rotor performs a similar function. Another approach is to induce a strong suction or low-pressure region *behind* the rotor. In a ducted, 10-ft diameter rotor, for example, the duct will extend perhaps 40 to 60 feet behind the rotor, expanding in diameter as it trails back. Shorter, more exotic concepts are now being tested. But when considering a duct, keep in mind that getting air to speed into a tube that has something resembling a cork inside is like getting speeding cars to penetrate a freeway roadblock. They would rather go around.

A different approach now being tested by several research teams is to mount airfoil vanes on the blade tips. These vanes help to expand the wake behind the rotor, causing more suction of air through the rotor. The real trick here is to get more wind power into the rotor than the vanes take out because of their added drag. So far, little success has been achieved in field tests of this method.

In evaluating these enhanced-performance methods, you should be careful that the extra cost is more than offset by increased power output. All too often, a 20-ft diameter duct around a 10-ft rotor performs worse than a 20-ft rotor that uses much less material for construction and less engineering time for design. However, many of the truly innovative improvements to wind systems are headed in the direction of enhanced-performance machines. There's still a lot of room for new inventions and fresh thinking.

Choosing a Suitable Wind Machine

Windmills vary in type, efficiency, and size; the choice you make in selecting a design type should be based on the nature of your project. If the design is to be extra low-cost, using local recycled materials, then you might select a design of low aerodynamic sophistication such as a Savonius rotor or one of the other drag-type derivatives. The choice will depend on the type of load you select. A water pump is one type of load; a generator is completely different. Keep in mind that rotor efficiency is almost directly proportional (with few exceptions) to the level of aerodynamic sophistication. Thus, your selection of rotor type will determine its efficiency.

Should you choose a sophisticated rotor (e.g., a three-bladed propellor-type rotor) the final system performance will be determined by how carefully you execute the design concept. If the three-bladed rotor is constructed of flat sheets of material such as plywood or sheet metal, expect overall performance to be low. Sail airfoils will actually improve performance. Twisted, tapered and carefully built airfoil blades made of metal or fiberglass will bring performance up to nearly optimum values.

To choose a suitable windmill, then, you must answer such fundamental questions as:

- Do I want a really cheap wind system?
- Am I willing to sacrifice efficiency for cost?
- Exactly what do I want the system to do?

Many people want to power their house—to build some sort of inexpensive machine that will replace the utility wires. Other folks want to supplement their power, lowering the monthly electric bill. But the central question remains: How much reliance can be put on the wind system for its power?

Efficiency, cost, and reliability are the key points to consider as you begin to plan your wind system. Just keep in mind that wind machines, like cars or airplanes, have certain historical cost limits. Above these limiting costs, the machinery becomes extravagant, fancy, or just plain expensive. Below these limits, the machinery may be skimpy in its design, unreliable, and—in the end—just as expensive. For wind machines, the optimum cost for the equipment seems to float between $500 and $1,000 per kilowatt of rated power. These values are close to what your electric utility pays for construction of a new coal-fired power plant.

If costs are lowered by skimping, scrounging, or using surplus machinery not well suited to the task, system reliability and performance are usually sacrificed. Using an old farm water-pumper to generate electricity by replacing the crankshaft with a gear- or chain-driven generator may seem like a quick way to shut off old Edison. But it won't work. This rotor is usually small in diameter—around 6 feet—and was designed for low rpm operation. A large gear ratio—from the slow-turning rotor shaft to a fast-turning generator—of about 50 to 1 will be needed. A 10 to 1 ratio might coax a few watts out of the generator, but overspeeding of the poorly loaded rotor blades will eventually encourage them to fly apart.

Choose a design for the task. Mechanical loads like piston water-pumps, washing machines, or piston compressors usually need high starting torque. Here, a recycled multiblade water-pumper or a Savonius rotor make sense. For loads like generators that do not "kick in" until they are spinning quite fast, choose a high-speed lifting rotor such as the two- or three-bladed propellor or a Darrieus rotor.

One overlapping design is the sail-wing rotor. Sew large sails and you have a high-torque, slow-turning rotor for water pumping. Sew smooth, narrow sails, and the rotor rpm increases—making electrical generation reasonable. The sail-wing rotor is also very appropriate for low-cost, low-technology systems constructed by owner-builders.

The final points you must consider in

Power Factor F	
V	**F**
6	1.07
7	1.76
8	2.62
9	3.74
10	5.13
11	6.82
12	8.86
13	11.26
14	14.07
15	17.30
16	21.00
17	25.19
18	29.90
19	35.17
20	41.02
21	47.48
22	54.59
23	62.38
24	70.88
25	80.11
26	90.12
27	100.92
28	112.55
29	125.05
30	138.43

Sizing a Wind Rotor

There are two principal ways to determine the frontal area of a wind machine rotor. You can merely guess how large a machine you want, calculate the power it produces, and stop there. Or you can first determine your average power needs and the wind resources at your site, and then equate the two to determine the rotor area. The first method is the one most often used. The second is more complex but results in a much closer match between your power needs and the wind power available.

Suppose you know in advance your average power needs—denoted here by the letter P. Equation 1 tells you that this power, if supplied by a wind machine, depends on the windspeed V, the rotor area A, the air density ρ, and the system efficiency E:

$$P = \tfrac{1}{2} \times \rho \times V^3 \times A \times E.$$

This formula can be rewritten to express the rotor area A in terms of five factors:

$$A = \frac{P}{E \times F \times C_A \times C_T} \,, \quad (Eq.\ 4)$$

where F is a factor that depends on windspeed and is presented in the first table here, and C_A and C_T are the altitude and temperature correction factors to the air density that are given in the tables on page 48 of Chapter 3. Equation 4 gives you the area in square feet when the power P is expressed in watts; if P is in horsepower, multiply A by 0.737.

If you are purchasing a factory-built machine and know its system efficiency, this formula can tell you whether its frontal area is suited to your power needs. If you intend to design and build your own machine, you need an estimate of the efficiency before you can begin. Use the second table here to get

Rapid Efficiency Estimator		
Wind System	**Efficiency, %**	
	Simple Construction	**Optimum Design**
Multibladed farm water pumper	10	30
Sailwing water pumper	10	25
Darrieus water pumper	15	30
Savonius windcharger	10	20
Small prop-type windcharger (up to 2 kW)	20	30
Medium prop-type windcharger (2 to 10 kW)	20	30
Large prop-type wind generator (over 10 kW)	—	30 to 45
Darrieus wind generator	15	35

a rapid but rough estimate. Then get values of F, C_A and C_T for your site from the appropriate tables. The rest is calculation.

Example: You have chosen a three-bladed propellor-type machine to produce 1000 watts. A site survey shows that the energy content of the winds at your site, which is at sea level, peaks at 15 mph. What size rotor is needed?

Solution: Begin by estimating the system efficiency. For small propellor-type systems, you can expect an efficiency from 15 to 30 percent. You elect to use 25 percent (E = 0.25) for a carefully designed machine. From the first table F = 17.30 at 15 mph, and C_A = C_T = 1 at sea level, for standard temperature (60°F). So,

$$A = \frac{1000}{0.25 \times 17.30 \times 1 \times 1}$$

$$= 231\ ft^2 \,.$$

You usually need to know the diameter of the rotor that can do the job. In this case, a diameter of 17.3 feet is needed. Refer to Appendix 3.1 for more detailed information needed to convert rotor frontal area into linear dimensions of the rotor vanes or blades.

choosing an appropriate design are based on real calculations of your power needs and system power output. These calculations can be done by using steps presented in the next chapter and in Appendix 3.3. Chapter 5 also presents a number of design methods you can use to select aerodynamic and structural components for your wind machine. In general, several methods are presented—from the simplified approach to a review of some of the more complex mathematical solutions. When combined with a careful analysis of your resource and needs, these methods allow you to design an appropriate wind system.

5
Wind Machine Design

Matching a Rotor to the Site and Task

Wind machine design is a process of trial and retrial. Once you have selected a wind-mill type, estimated the overall system efficiency and calculated rotor size, you must then design the components of the machine. With all components sorted out, you next evaluate in more detail the various efficiencies and performance characteristics, and re-estimate the system efficiency. Then you recalculate rotor size from this information, and if different from your first try, you redesign the components. With a bit of luck and fresh batteries in your calculator, you won't need many retrials to achieve an acceptable design.

The wind machine design process consists of two major tasks: aerodynamic design and structural design. Although these two tasks are logically separate, it helps to think of them together. Otherwise, you might find yourself laying out a bit of aerodynamic glory that just won't hold together in real life. Start by identifying the load that the rotor must power. If the load is a mechanical device—e.g., a water pump—high starting torque from the rotor will be needed. This will usually be true for a compressor as well. If the load is an electrical generator, low starting torque but high rpm will be required.

The starting torque required tells you the necessary rotor solidity—the ratio of total blade area to rotor frontal area. High torque needs high solidity. Low torque means low solidity. You choose rotor solidity according to the type of load.

The solidity tells you how much surface area the blades must have. For example, if the solidity equals 0.2, and the rotor is to be 79 square feet in frontal area (about 10 feet in diameter) then the total blade area equals 0.2 times 79, or 15.8 square feet. Solidity also tells you something about the rpm and the tip-speed ratio at which the blades will operate. High solidity means low rpm and low TSR, while low solidity means the rotor must travel faster—high rpm and high TSR. Putting a design together, then, starts with a statement about the task the machine must perform and the rotor load that task creates. These considerations lead naturally to some very specific constraints on the rotor geometry and operating characteristics.

Aerodynamic Design

Sorting out airflow, airfoil selection, blade twist, torque and performance coefficients—that's aerodynamic design. The sophistication required is really determined by two things: size of the wind machine, and tip-speed ratio. Small wind machines can be built "like the picture"; they can be aero-dynamically shaped with only a little care. Performance will then be a matter of luck. Aerodynamic perfection becomes increasingly important for wind machines larger than 2 kW. With large machines, experimenters cast aside ideas of giant Savonius rotors or funky sail-wing machines and start to get serious about rotor design.

In the low tip-speed range (less than

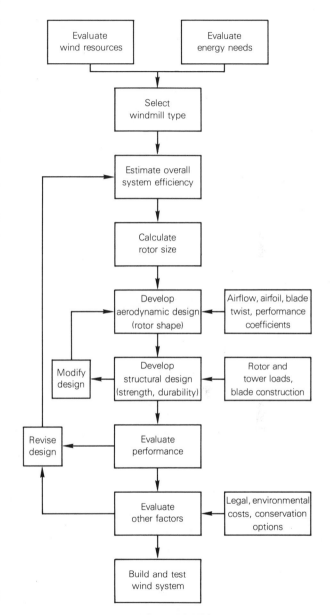

The wind machine design process.

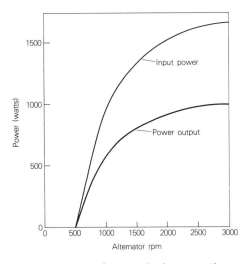

Power curves for a typical automotive alternator. A constant 60 percent conversion efficiency has been assumed.

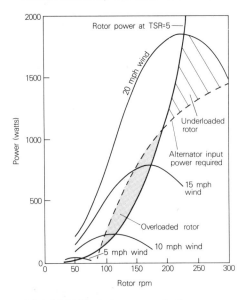

Matching the rotor power to the power required by an alternator.

TSR = 8) rotor design is governed less by aerodynamic considerations than one might expect. A fully optimized rotor blade will perform perhaps 5 percent better than a carefully designed blade that was compromised somewhat to lower the cost or allow for easier construction. The extra power available may not be worth the added time or expense. But a fully optimized blade makes more and more sense above a TSR of 8.

Throughout the process of aerodynamic design, you will be concerned with the following factors and the way they relate to each other:

- Torque, power, and rpm requirements of the load
- Torque and power characteristics of the rotor
- Response of the rotor to gustiness in the wind.

Just how concerned you must be with these factors depends again on the machine size. For example, a wind machine that supplies power to the utility grid will be large and expensive. A very detailed understanding of this machine's response to gusts, or its match with the load, is crucial to the final economic success of such a system. A small windcharger, on the other hand, might not have a well-matched rotor and load, but could be an acceptable machine because of its lower cost. Still, it's a good idea to pay attention to such factors in any project, regardless of size.

For a specific rotor, you can calculate a power curve that shows rotor power output versus windspeed or rotor rpm. You can also plot a power curve versus rpm for the load you choose. For example, power curves are available for most electrical generators, and rotor power can be calculated using Equation 1 and an estimate of rotor efficiency. You might end up with a rotor that turns at 300 rpm at its rated windspeed, while the generator needs to turn at 600 rpm to generate any power at all. Hence, a transmission is needed to match rotor rpm to that of the load.

The two graphs at left present typical power curves for a small wind generator with a 12-foot rotor driving a 1-kW, 12-volt alternator. The first gives the alternator input and output power curves available from the manufacturer. The other shows the rotor power curve with the alternator curve superimposed, after including the step-up in rpm from the transmission. The transmission, which could be a belt and pulleys, or a gear box, increases rotor rpm to the much higher rpm required by the alternator.

By comparing the rotor power curves with the alternator power curve, you can tell something about rotor performance at various windspeeds. Notice that the alternator places virtually no load on the rotor at a windspeed of 5 mph. That's fine because very little wind power is available here. But by 8 mph (in the case of a 6:1 gear ratio) the alternator is beginning to demand higher power input from the rotor than is available at the design TSR. A 6:1 gear ratio mildly overloads the rotor up to about 16 mph.

Above that windspeed the rotor is underloaded—more power is available than the alternator can use. A highly overloaded rotor may stall and quit turning altogether. On the other hand, an underload condition will cause the rotor to speed at higher rpm—higher than the optimum tip-speed ratio. This example is typical of many low-power wind generators using available automotive alternators.

In small design projects, the process of load matching is often reduced to selecting the transmission gear ratio that minimizes the effects of mismatch. For certain types of pumps and compressors, a gear ratio can be selected that almost ideally matches rotor to load. For alternators and generators, however, some amount of load mismatch will occur.

How the rotor performs when the windspeed changes abruptly is another important design consideration. The diagram shows a typical wind gust that nearly doubles the windspeed in a few seconds. For clarity, the windspeed is illustrated as staying at the new speed, a very unlikely occurrence. Two typical rotors respond to this gust by accelerating to higher rpm. One rotor is lightweight (low inertia), perhaps a Savonius rotor made of aluminum; the other is heavy (high inertia), perhaps the same size S-rotor, but made of steel drums.

Notice that the heavy rotor accelerates more slowly than the light one. Really large rotors might take half a minute to follow a gust, which disappears before that time.

Heavy rotors tend to average the windspeed—staying at an average rpm. Light rotors also average, but they experience more fluctuations in rpm. The significance of this averaging effect is that a rotor cannot always operate at its optimum tip-speed ratio. It will operate at an average TSR, yielding less than maximum efficiency.

The shape of the curve depicting the relationship of rotor efficiency to TSR is very important in the overall performance of that rotor. Two different efficiency curves are shown in the next diagram; the dashed curve shows a higher maximum efficiency than the solid curve. But the solid curve is broader and flatter, so a gust-induced change in TSR would produce a much smaller change in rotor efficiency. Hence, because of the averaging tendency of rotors, a flatter efficiency curve is often more desirable than a peaked curve—even if its maximum is slightly lower. Over the long run, the machine with a broad, flat efficiency curve will generate more wind energy than a machine with a sharply peaked curve.

Savonius Rotor Design

Aerodynamic design of the Savonius rotor and other simple drag machines is mostly a matter of drawing something that looks like it will work. For most Savonius projects, the shape is determined less by design factors and more by available materials. The usual "home-brew" rotor is made

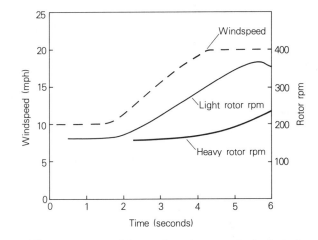

The response of two rotors to a gust of wind. A light rotor speeds up more quickly than a heavy one.

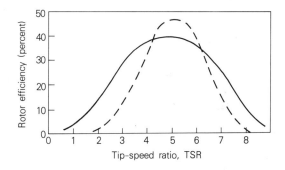

Performance curves for two small wind machines. A broader curve is the more desirable, even though peak efficiency may be slightly reduced.

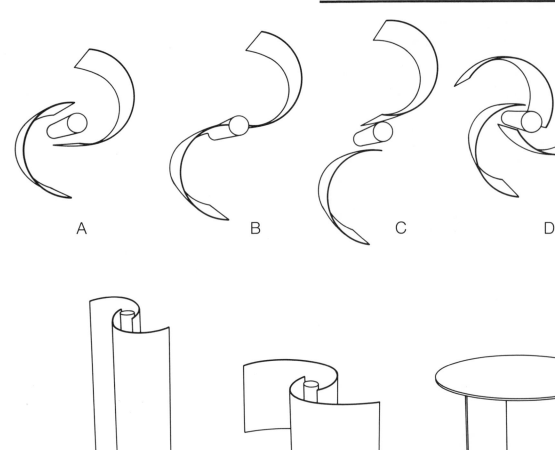

A   B   C   D

E   F   G

Savonius rotor design options include the intervane gap, number of vanes, aspect ratio, and tip plates. Option E has a much higher aspect ratio than F, and the tip plates in option G improve the rotor performance at low rpm.

from oil drums cut in half; an occasional rotor is built from sheet metal. Either the drum size or sheet metal width will determine rotor diameter. Most S-rotors are about 3 feet in diameter, but some have been built 30 feet across.

Many studies have been conducted to determine optimum shapes for Savonius rotors. The results are summarized in the diagram on this page. Options start with the *intervane gap.* In the diagram, option A is an improvement over options B and C. Air can flow through the intervane gap in design A and push on the upwind-traveling vane, reducing the drag on this vane and increasing torque and power. The number of vanes is the next consideration, and A seems to be an improvement over D. Theoretical explanations for this effect are complex and possibly incorrect. But experimental tests show that two vanes work the best. Because of the materials you have available, such as old oil drums with which to make an S-rotor, the number of vanes and the intervane gap might be limited by your ability to fit the pieces together. There is not an enormous difference in performance between the various options, but better performance is possible when you can use the best options.

The next design variable is the vane *aspect ratio*—in this case, the ratio of vane height to diameter. There is probably no best design in this case. For a given frontal area, higher aspect ratio rotors will run at higher rpm and lower torque than those with a low aspect ratio. Tip plates improve S-rotor performance slightly, especially at very low start-up rpm.

What kind of performance can you expect from an S-rotor? The graph here presents typical performance curves for two S-rotors with different vane gaps. The best one shows a maximum efficiency of about 15 percent—a typical value for a small machine like an oil-drum S-rotor with a 3-foot diameter. For a machine with a diameter of 10 feet or more, you could expect an efficiency of 20 percent. Note that the graph shows both torque and power coefficients and illustrates the S-rotor's characteristically high starting torque. Loads driven by a Savonius should have roughly similar torque and power requirements. See Appendix 3.2 for a sample design calculation that uses the Savonius performance curves presented here.

Propellor-Type Rotor Design

If you wish to use a wind machine to drive an electrical generator, the high rpm needed by the generator will require a high-speed rotor. In general, only propellor-type and Darrieus rotors can develop the high rpm needed. In both cases, careful aerodynamic design of the rotor blades is important if maximum rotor efficiency is desired.

Earlier, you saw how tip-speed ratio and rotor efficiency are closely related. Generally, machines that operate at higher TSR have higher rotor efficiencies—as long as they are not overspeeding. The tip-speed ratio for a specific rotor is not derived by guess-

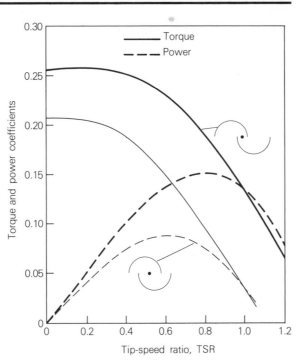

Performance curves for two S-rotors. Use these curves to estimate rotor torque and power.

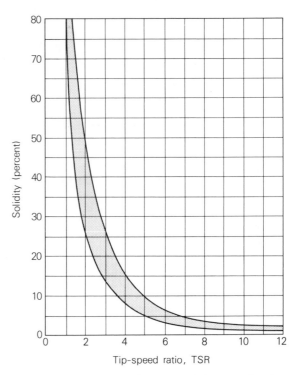

Rotor solidity as a function of tip-speed ratio. Use this graph to estimate blade size.

work; it is governed instead by the blade size. Wide, large-area blades like those in a farm windmill turn at low TSR; long, slender blades spin at high TSR. There is thus an inverse relationship between solidity and rotor speed or TSR. An American Farm multi-blade windmill has high solidity—about 70–80 percent of the frontal area is covered with blades—and operates at low TSR. But, as seen in the above right graph, high-speed, propellor-type rotors tend toward low solidity—10 percent or less—and high TSR. Compare the graph with the table.

BLADE NUMBER VS. TSR	
Tip-Speed Ratio	Number of Blades
1	6-20
2	4-12
3	3-8
4	3-5
5-8	2-4
8-15	1-2

HORIZONTAL-AXIS WIND MACHINES			
Wind Machine	Design TSR	Blade Type	Blade L/D
Water-pumper	1	Flat Plate	10
	1	Curved Plate	20-40
	1	Sail-wing	10-25
Small wind generator	3-4	Simple Airfoil	10-50
	4-6	Twisted Airfoil	20-100
	3-5	Sail-wing	20-30
Large wind generator	5-15	Twisted Airfoil	20-100

Sizing the Rotor Blades

For a specified number of blades, the solidity of a rotor tells you approximately how much surface area each blade has. You can calculate this area from the following formula:

Total Blade Area = Solidity × Rotor Area.

Formulas for the rotor area, or frontal area, of Savonius, Darrieus, and propellor-type rotors are presented in Appendix 3.1. To get the surface area of a single blade, just divide the total blade area by the number of blades.

For example, suppose you need a TSR of 5 and have selected a three-bladed, propellor-type rotor with a diameter of 12 feet. From the graph of solidity versus TSR, the maximum recommended solidity at TSR = 5 is 0.1. A 12-foot diameter rotor has a frontal area A = 113 square feet. Then

$$Total\ Blade\ Area = 0.1 \times 113$$
$$= 11.3\ ft^2 .$$

The surface area of each blade in this three-bladed rotor is then 3.77 ft².

Note that solidity also helps to determine the number of blades in a complete rotor.

After selecting an appropriate solidity for the load and operating TSR you have in mind (the "design TSR"), you need to know something about the blade types that will deliver these hypothetical values. The design TSR is the tip-speed ratio at which all of the aerodynamic factors are close to their optimum values and the rotor efficiency is at its maximum. The table at left provides this information for three types of horizontal-axis wind machines. It also provides estimates of the blade lift-to-drag ratio L/D that will be needed for further analysis.

Keep in mind that different tip-speed ratios imply different values of the solidity for the rotor you are choosing or designing. Solidity plays an important role in two different design areas:

- Rotor torque characteristics, especially starting torque
- Rotor weight and cost owing to materials usage.

Low solidity means, generally, less materials because of the smaller blade area. A trade-off exists between rotor solidity and cost per area. Long, thin, high-performance blades usually require much more material *inside* them for structural support than does a thicker, fatter airfoil. You want to maximize performance per dollar, but both performance and cost decrease with increasing solidity. Each wind machine you design can be subjected to a computerized study of such trade-offs. In the example (see box), we use TSR = 5 to calculate blade size with the solidity diagram. The 12-foot diameter rotor puts this small but useful machine in the 2-kW class. Such a low-power windcharger needs two to four airfoil blades, which can be carved wood, molded composite, or metal. Notice that the lift-to-drag ratio for these blades can be anywhere from 20 to 100. The blade L/D is bound to be less than the ideal airfoil L/D because of imperfections in manufacturing, aerodynamic pressure losses at the blade tips, and other similar factors. But a rotor operating at TSR = 5 needs a blade with an L/D of 50 or more for optimum performance.

Another way to visualize the blade L/D requirement is illustrated in the diagram at right. Notice that operation at higher TSR requires higher blade L/D values to maintain good performance. Your TSR selection is limited by your estimate of how well you can build a set of high-performance blades. There is no sense in selecting a TSR of 10 if the blade you fabricate has an L/D of only 20. In an aerospace factory environment, tooling engineers design assembly fixtures that permit extremely accurate reproduction of computer-generated blade designs; high lift-to-drag ratios are thus attainable there. Small wind systems built with minimum tooling and no computer simulation will probably have lower lift-to-drag ratios. To lower the sensitivity of rotor performance to blade L/D, select a lower design TSR. Then assess the blade weight and cost that result from the higher solidity.

To take the analysis one step further, the nature of the wind machine must be examined. Basically, there are two classes of high-speed rotors:

1. Rotors designed to operate at a constant, optimum TSR; the rpm will vary with windspeed. These rotors are typically found on small machines—driving air compressors and direct current electrical generators.
2. Rotors designed to operate at a constant rpm; the TSR will vary with windspeed. Such rotors are usually designed for large, synchronous alternating current generators whose constant output frequency depends on the rpm remaining constant.

For Class 1 rotors, a design TSR can be selected in the range of TSRs appropriate to each wind machine and blade type. For Class 2 rotors, you must derive a range of operating TSRs from the design rpm of the generator, the transmission gear ratio, and the range of windspeeds that provide most of the energy at the planned site. (See the section on windspeed distributions in Chapter 3.) The operating TSRs should fall within the range indicated for high-power wind generators, and the corresponding lift-to-drag ratios will apply to blade design and manufacture.

In the example of a rotor-to-generator load match presented at the start of this chapter, you saw that such inexpensive loads as automotive alternators don't completely match the power curves of a rotor.

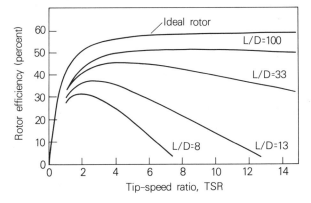

Blade lift-to-drag ratios L/D versus rotor efficiency. High L/D's are needed at high TSR.

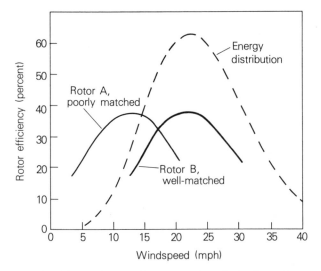

Matching rotor performance to wind characteristics of a site. A well-matched rotor follows local energy patterns closely.

Such a mismatch will slightly distort the rotor efficiency curve, which will peak at about the same windspeed as the generator output if the generator and rotor are matched with no overload or underload. An optimum windspeed actually exists for such a windcharger, and such a windspeed should match the peak of the energy curve for winds at a given site. Such matching involves changing transmission gear ratios and voltage-regulator characteristics.

In a production situation, it is far cheaper to reduce the overload/underload condition as much as possible by careful selection of a single gear ratio and to produce that product for most wind sites. Changing gear ratios for site optimization does not pay off in small wind machines—as long as care is taken with initial rotor-to-load matching.

In the example discussed, the tip-speed ratio is predetermined; the rotor has already been designed. But suppose you want to select a different TSR based on other reasons, such as:

• Higher TSR means a lower gear ratio is required

• Flutter and vibration problems (with very large, slender blades) limit the rotor to a certain maximum rpm.

If the project is small and the rotor is stiff (that is, it does not flex easily) and does not experience vibration problems, then a higher TSR might be considered. But for wind rotors with diameters larger than 30 feet, vibration, flutter, and other problems associated with blade flexing become important design

considerations. They are discussed later in this chapter.

Once an appropriate solidity and TSR have been selected, what should the blades look like? Should they be tapered, straight, fat or thin? How you plan to manufacture the blades will help determine blade design. For example, extruded airfoils cannot be tapered along their length; they must have the same profile from root to tip. The tip chord length (the distance from leading edge to trailing edge of the airfoil) has to equal the airfoil chord at all points along the blade radius.

How is blade chord length calculated? The graphical method of blade design in Appendix 3.3 will allow you to calculate the blade chord lengths needed along the blade span and the airfoil angles at each step. Using this blade design method will give you a blade both tapered and twisted. The chord will be much shorter at the tip than at the root, where the airfoil will have a much steeper angle than at the tip. This twist will not be linear; there will be gentle twist at the tip and lots of twist near the root.

But this simplified analysis does not allow you to evaluate the rotor efficiency, or C_p. How is an accurate rotor efficiency curve generated? Appendix 3.3 also references a slightly different blade-design procedure, presented in a report by Peter Lissaman and Robert Wilson. Their report contains a computer program that calculates virtually all the performance curves you need.

Certain simplifications in blade design

are possible on small wind machines if you are willing to sacrifice a little performance. The graphical design method presented in Appendix 3 was used to design the blades on the Kedco wind generator. A decision to simplify the blade construction required that nonlinear twist and taper be changed to linear twist and linear taper. That way, the leading and trailing edges are straight lines connecting tip and root. The actual twist of the blade was determined by the amount of twist possible in sheet aluminum skin without wrinkling that skin.

These simplifications resulted in an easy-to-build blade with a maximum rotor efficiency of 42 percent. Under ideal conditions, you would not expect an efficiency much higher than 45 percent for this size blade. So the penalty paid for ease of construction was just a few percent. But remember that the operating TSR of this Kedco wind generator is 4 to 5. If you design for higher values, the performance penalty will increase.

Darrieus Rotor Design

Much design work has gone into perfecting the Darrieus rotor. As it turns out, the aerodynamics of the Darrieus let you use the graph of solidity versus TSR to design rotor blades. Start by sizing the rotor using an estimated system efficiency from the last table in Chapter 4. From the output power required, windspeed, and this efficiency estimate, you calculate frontal area required.

The Kedco wind generator uses simple blade design to achieve a 42 percent peak rotor efficiency.

then determines the solidity, as well as gear ratios, generator speeds, and structural design of the rotor. Using this TSR and the graph on page 95, select a value of the solidity. As with the prop-type rotor, the solidity allows calculation of blade area: solidity times the rotor frontal area equals total blade area. Divide the total blade area by the number of blades (usually 2 or 3) and you get the individual blade area. Divide this individual blade area by the rotor height to get the chord length. In Darrieus rotor design, the rotor height is used in much the same way as blade length or rotor radius is for propellor design.

Airfoil selection for a Darrieus rotor is limited to symmetrical profiles. Symmetrical airfoils are used in the Darrieus because lift must be produced from both sides of the blade. On large Darrieus rotors with long chord lengths, you can have a blade profile that is symmetrical about the curved path that the rotor blades trace. Such an airfoil will look cambered, but it will perform as a true symmetrical airfoil while speeding along this path.

Extruded metal blades for a Darrieus rotor. Symmetrical blade profiles are needed because lift must be produced on both sides of the blade.

Next, the linear dimensions needed to give that frontal area are calculated by referring to Appendix 3.1. For both the eggbeater Darrieus and the straight-blade design, assume that height equals diameter.

The operating tip-speed ratio for a Darrieus lies between 4 and 6. This design TSR

Structural Design

Now that the rotor size and shape has been determined, will it stay together? If it spins too fast, can it withstand the centrifugal forces on it? If a sudden gust of wind slams into the machine when it is not spinning, will the blades bend and break off?

These and other questions must be answered during the structural design process. In addition to meeting its performance goals, the machine must stay together.

Structural design is a two-step process:

1. Start by examining the various forces and loads on the rotor and its support tower. Determine which forces gang up or act together on each structural member (blade, powershaft, etc.).

2. Design each structural member to withstand the applied load. In a wind machine, this means designing the member to withstand *static loads* without bending or breaking and to take repeated applications of those loads without failing from fatigue.

Furthermore, blades also need to be stiff enough to prevent *flutter*—resonant flexing oscillations that can fatigue the blade rapidly.

Starting from the ground up, the first load encountered by a wind system is its often tremendous weight. At ground level, this weight acts against the foundation, which must prevent sinking. Drag, or the force of the wind acting on the wind machine and tower, is also encountered. The tower foundation must prevent the tower from toppling from this drag force. Torsion, or twisting of the tower and its foundation, is caused by *yawing,* or a wind machine changing direction with respect to the tower axis. Most small wind machines are free to yaw; for them, this torque load does not exist. Very large wind machines with servomotors to control the yaw of the rotor will apply torque loads to the tower. A major load seen from the ground is applied when the rotor is unbalanced. This load is similar to the gravity load (weight) but is oscillatory (up and down) or torsional, or both. It is not "static," or continuously applied, but dynamic—and often catastrophic.

The tower might be free-standing, or cantilevered, or it may be braced with struts or guy wires. It may even be the roof of a house, in which case one should be awfully careful about those occasional dynamic loads mentioned above. Otherwise, a mere catastrophe can quickly become a disaster!

Moving up the tower and into the wind machine, an appropriate load to start with is the torque load in each of the power shafts. One power shaft—the main shaft—takes rotor power into the transmission. The other power shaft takes power out of the transmission. No transmission means you have only one power shaft and one torsional load to examine.

Depending on how the machinery is designed, part of the rotor weight might cause the main power shaft to bend. Shaft bending is different from beam bending, as in bridges, because the shaft is spinning. When combined with torque, shaft bending can cause severe problems. Any bending deflection in the shaft would show up as a nasty whirl in the rotor. Bending can also show up in the support frame—often called a *bed-plate,* or *carriage.* If rotor loads are taken through the rotor bearing into the carriage instead of the power shaft, few if

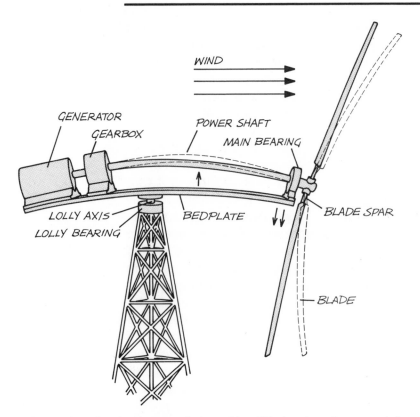

WIND

GENERATOR

GEARBOX

POWER SHAFT

MAIN BEARING

LOLLY AXIS

LOLLY BEARING

BEDPLATE

BLADE SPAR

BLADE

Various bending loads on a wind machine. Wind and gravity cause static loads; dynamic loads are caused by variations in windspeed.

the inertia of all the machinery at one end of the carriage strains against the yaw driver at the other end. The yaw driver might be the rotor (in the case of a downwind rotor), or a rudder or tail vane (if the rotor is mounted upwind of the tower). Miscellaneous loads on the carriage are associated with mechanics sitting on top of the generator (always add about 200 pounds to the generator weight to compensate!), or Mr. Goodwrench dropping the transmission into place. Keep in mind that hefting 100 pounds or more of iron 80 feet in the air is no simple task.

The final loads to examine are those affecting the rotor itself. Start with the rotor at rest, or "parked." No wind. The first load is the blade weight. For small machines, don't worry about it, unless you're installing the rotor atop an erect tower. For large machines, blade-bending loads under static conditions can approach the design load. Long blades are enormously heavy. Such a load can easily buckle a thin, streamlined blade. Add some wind, and aerodynamic drag combines with weight—the static duo and a nasty set of loads to contend with. Park the blades vertically, and the gravity load loses its importance. The bending due to drag does not, however.

Blade bending during operation is caused by two forces at once: lift-induced thrust—acting in the plane of rotation—and drag-induced bending that acts downwind. These two vectors combine to deliver one huge force along the blade span, and the blade had better be able to take it. The worst case

any problems will beset the power shaft.

The carriage must bear all of the static weight of the components (generator, transmission, rotor), and any dynamic or shaking loads from the rotor. In machines that use servomotors to aim the rotor into the wind, the carriage must also bear the static and dynamic loads that occur in the lateral direction—parallel to the earth's surface. Smaller lateral loads appear in free-yaw machines;

of blade bending occurs because of the governor. This load can be difficult to analyze. If the governor changes the blade angle slowly, the blades and other rotating machinery will slow down together, gently. If the governor changes the blade angle abruptly, the rotating machinery will, because of its inertia, try to keep the rotor turning at a time when its aerodynamic loads have diminished. The result is blade bending induced from the power shaft. Depending on the inertia of the rotor, this load may be negligible or very large, and it occurs in the reverse direction—contributing to blade fatigue problems.

Some rotors are controlled or stopped by a brake mechanism that can exert a load like a governor. This load must be carefully controlled to avoid severe bending. A similar load occurs from a malfunctioning generator or gear box, or a water lock in a pump. Suppose the rotor is spinning in a good wind with only a small generator load applied when the voltage regulator suddenly kicks in a full load. Depending on the strength of the generator, the result will range from unimportant to noisy. It will also contribute to fatigue in the blade's structure. If a water lock occurs in a pumper, as it often does, the result could be anything from a buckled sucker rod to a pretzel-shaped rotor.

Another rotor load is the centrifugal load, or centrifugal tension, in the blade structure. The blade spar is connected to the hub— either through a feathering bearing, or directly. This connection is an area where many a blade has parted company from the rest of the rotor. The result is always holocaust, anger, and dead batteries. The centrifugal tension load results from the mass of the blade spinning about the powershaft. Under certain conditions, centrifugal tension can be used to reduce blade bending—centrifugal stiffening as it's called. Centrifugal tension, then, can be useful and is not the worst load, except in the case of a runaway rotor. This load increases much faster than the rpm, and a runaway rotor can easily spin twice as fast as its maximum governed rpm, causing four or more times the maximum centrifugal tension in the blade. Take heed. Broken blades can fly a long way. In the case of Darrieus rotors and other vertical-axis machines, blade bending is also caused by centrifugal force due to rotation. Particularly in the straight-bladed Darrieus, these centrifugal blade-bending forces can be awesome. One solution to this problem is the use of cable supports that extend from the blades to the axis of rotation.

The final rotor load to worry about is torsion within the blade itself, a twist along the blade axis. The "tennis-racket effect" says that a spinning blade should lie flat in its plane of rotation, but twist and other design considerations set the blade otherwise. Also, airfoils other than the symmetrical variety have a pitching moment that applies a torsional load into the blade. These are the two major causes of torsion within a blade.

Unfortunately, structural design does not stop with sorting out the loads just listed. A host of environmental considerations send designers up the wall. Hail, dust, rain, and sleet work to erode or destroy a rotor. Leading-edge protection may be needed to keep blade airfoils from disintegrating. Migratory bird impacts have taken their toll of both rotors and birds. By far the worst, though, is gunshot damage inflicted by passing hunters. No ready design solutions are yet at hand, but it's possible to get angry enough to imagine laser-targeted firing mechanisms that permit the wind machine to shoot back.

Lightning or simple electrostatic build-up and discharge ought to be considered during structural design. Certain types of composite materials, like carbon fibers, turn into frazzled, fuzzy, nonstructural messes when zapped by lightning. Other materials, including fiberglass, behave admirably well, considering the amount of electrical energy transmitted in a direct hit.

Many loads will require careful engineering analysis using books containing structural design data and engineering expertise. To pass some building department inspections, you may need the services of a licensed engineer. The rest of this chapter contains a discussion of some of the calculations required to complete a successful preliminary design for a large project or a finished design for a small machine. A benefit of buying a machine off the shelf is that most structural considerations have already been engineered into it. The customer pays only part of the engineering fees, not all.

In view of the safety aspects of a wind machine installation, you should realize that this chapter is no substitute for sound, rigorous engineering design. The discussion and methods are simplified so that a large segment of readers can appreciate the nature of the design requirements.

Estimation is the first step in determining the various loads. For example, to estimate the weight of your machine, find out what similar machines weigh. Generally, small machines weigh from 100 to 300 pounds per kilowatt of rated power. You must guess where yours will fall within a known range. Once the machine is fully designed, you can calculate rather than estimate the actual weight of each component. Transmission and gearbox weight can be obtained from the catalogs available from many manufacturers of such components.

A third and final procedure is to revise some of the load and structural design calculations you made on the basis of earlier weight estimates. Blade loads are a prime example of this procedure: Estimate, calculate, then revise. During this design procedure, you should be asking such questions as: Is the design easily buildable? What materials should be used? Is this the lowest cost alternative? Will the design be easy to maintain? Will it require a lot of maintenance?

Blade Loads

Now let's consider the loads on individual rotor blades. First you estimate the weight of a single blade, say 10 pounds. Then, from your predicted operating TSR, you calculate maximum rpm. Knowing blade weight and rpm, you can now calculate centrifugal tension in the blade under normal operating conditions. You can also calculate blade-bending loads. Using these, you can design the blade to withstand these loads and calculate its weight as designed. Compare estimated weights to calculated weights and correct the calculated loads accordingly. The idea is to converge the loads and weights to their final design values.

The first blade load to consider is centrifugal force—the result of heavy objects moving rapidly in a circular path. If given the chance, such objects would rather travel in a straight line. Tie a rock to a long string and swing the rock around while holding on to the string. Centrifugal force keeps the string straight and tight. Swing it around fast enough and the string breaks. The strength of the blade spar or other support that holds a windmill blade in its circular path had better be greater than the maximum centrifugal force.

The information you need in order to calculate centrifugal force on anything moving around in a circle is the speed of the object, its weight, and the radius of rotation measured from the rotating object's center of gravity. If the object happens to be a windmill blade, its center of gravity can be determined by balancing the blade over the edge of a thin board, or by estimating its position during design. The method presented

Centrifugal Force

The centrifugal force pulling a blade away from the rotor hub is given by the formula:

$$Centrifugal\ Force = \frac{0.067 \times W \times (SR \times V)^2}{RC},$$

where

W = the weight of the blade in pounds,
SR = the speed ratio at the blade center of gravity,
V = the windspeed in mph,
RC = distance in feet from the center of rotation to the blade center of gravity.

Notice that this force is proportional to the square of the windspeed, so that it quadruples if the windspeed doubles and all the other numbers stay the same.

Problem: A three-bladed windmill with a diameter of 12 feet is designed to operate at TSR = 6. Each blade weighs 5 pounds with its center of gravity, as determined from balance tests, lying 3 feet from the center of rotation. What is the centrifugal force on this blade in a 10 mph wind?

Solution: First you need the speed ratio at the blade center of gravity, which is halfway out from the center of rotation to the blade tip. Thus SR = ½ × TSR = 3. Then,

$$Centrifugal\ Force = \frac{0.067 \times 5 \times (3 \times 10)^2}{3}$$
$$= 100.5\ pounds.$$

The force trying to rip the blade away from the hub is about 100 pounds in 10 mph winds. This may not seem like much, and it isn't. But repeat this calculation at V = 20 mph and V = 30 mph. You'll find that the centrifugal force is about 400 and 900 pounds, respectively. For a rotor that operates at constant TSR, centrifugal force increases as the square of the windspeed.

There are two operating conditions you should consider when calculating centrifugal force. First is the normal operating condition that occurs in maximum design windspeed just before the governor begins to limit the rotor rpm. The second is the abnormal, or runaway operation caused by a faulty governor. For a small machine, a 50 to 100 percent overspeed is not unreasonable. That is, a small rotor designed to operate at 300 rpm might hit 600 rpm in runaway condition. As a matter of fact, it might go even higher if it held together long enough. To estimate the centrifugal force under runaway conditions, use a speed ratio (SR) up to twice the normal design value in the equation above.

above will help you calculate centrifugal force on a rotor blade. This centrifugal force calculation gives you the information you need to select blade attachment bolts and other hardware. Also, centrifugal force is one of the loads you will use to design the structural members of each blade.

Rotor Drag and Blade Bending Moment

An approximate formula for the drag force on a windmill rotating in windspeed V is:

$$Rotor\ Drag = 0.0026 \times A \times V^2,$$

where A is the rotor frontal area in square feet and V is the windspeed in mph.
   Problem: *A three-bladed, propellor-type rotor 12 feet in diameter is generating its rated power in a 20 mph wind. What is the rotor drag on this machine?*
   Solution: *The frontal area of a 12-foot propellor-type rotor is 113 ft². Thus,*

$$Rotor\ Drag = 0.0026 \times 113 \times 20^2$$
$$= 117.5\ pounds.$$

Since the rotor has three blades, the drag force on each blade is one-third of 117.5 pounds, or 39.2 pounds. This is the drag force trying to bend the blade in the down-wind direction. It should not be confused with the aerodynamic drag on each blade that acts in the plane of rotation, trying to slow the rotary motion of each blade.
   The drag force on a windmill that is not rotating, or parked, is approximately:

$$Static\ Drag = 2 \times S \times Rotor\ Drag$$

where S is the solidity of the rotor, and the Rotor Drag is calculated from the first equation above. If the rotor in the example above

had a solidity of 0.1, for example, the static drag on each blade is:

$$Static\ Drag = 2 \times 0.1 \times 39.2$$
$$= 7.8\ pounds,$$

and the static drag on the entire rotor is 23.5 pounds.
   The blade bending moments, in inch-pounds, is calculated from the drag force on each blade according to the formula:

$$Bending\ Moment = 12 \times RC \times Drag,$$

where Drag is calculated for either the static or rotating conditions, as above, and RC is the distance in feet from the center of rotation to the blade center of gravity. This formula yields only an approximate value of the bending moment, but it is very close to an exact value that could be calculated from much more complex equations.
   Using this equation and the blade drag of the 12-foot diameter rotor already calculated, the blade bending moment is 1410 inch-pounds in the operating mode and 282 inch-pounds in the parked mode, if the center of gravity lies halfway between the tip and the hub of each blade. The point of maximum stress occurs right where the blade attaches to the hub—at the root of each blade.

Blade-bending loads are caused directly by lift and drag on the rotor. Rotor or windmill drag depends on rotor solidity and rpm, and the windspeed. Rotor lift contributes to bending in two directions: bending along

the direction of rotation and downwind bending. The blade structure must be capable of withstanding the total bending load. Rotor drag causes support towers to topple, guy wires to break, and rotor blades to bend. Consider the high-solidity, multiblade water-pumping windmills. Stand in front of such a machine, and it will resemble a solid disk, with no place for wind to flow through. The higher the solidity, the higher the rotor drag. In the case of a rotating windmill, drag is determined by approximately how much power you extract from the wind. A simplified formula for calculating the approximate rotor drag is presented here.

Conditions you should consider in this calculation are the rotating (operating) and static (parked) loads for normal winds, and for the highest winds expected at the site. In normal winds, you can expect that, because of governor mischief, your rotor will occasionally develop more than its rated power. As much as 50 percent more is a reasonable design value. Use the high power value and the windspeed at which the rotor will develop this power as design conditions. For the rotor, calculate drag at highest possible windspeed—more than 100 mph in many cases. Then, select the highest of these two calculations—parked versus operating—as representative of the loads applied to the tower and windmill structure. These drag loads will be used to calculate blade-bending moment by the procedure given here (see box), and tower bending moment, or guy-wire loads, as discussed later.

While rotor drag is used to calculate blade bending in the downwind direction, powershaft torque can be used to calculate blade bending in its thrust direction. Torque is not really a force, but more precisely the result of a force applied at a certain radius from the center of rotation—a twisting force. Blade lift causes powershaft torque that turns the shaft against the generator load. Torque is directly related to blade bending—mostly at the root end; it tries to twist the powershaft and shear the blade hub off it.

When the total torque has been calculated, each blade can be considered to contribute to it. In a three-bladed design, for example, one third of the torque represents the blade-bending load on each blade. Also use the total torque value to design the powershaft. This shaft must transmit torque loads between the rotor and its load without twisting along its length and failing.

Blade Construction

With a good estimate of the important blade loads, you next determine the blade strength—its ability to withstand those loads without breaking. The structural design of a blade and the materials with which it is made determine blade strength. The blade must be designed to withstand centrifugal tension, blade bending, and torsion. It must retain the airfoil shape and twist, and remain firmly attached to the hub. There are several ways to accomplish all of these design tasks.

Powershaft Torque

The powershaft torque, in inch-pounds, on a wind machine can be estimated from the formula:

$$Torque = 2245 \times \frac{D \times HP}{V \times TSR}$$

where

D = the rotor diameter in feet,
TSR = the tip-speed ratio,
V = the windspeed in mph, and
HP = the horsepower generated by the rotor.

Remember to use the horsepower output of the rotor, not the generator or other device being driven by the powershaft. If your evaluation of power is expressed in watts instead of horsepower, use 3.01 instead of 2245 in this formula to get a result in inch-pounds.

Problem: Suppose a 12-foot diameter windmill generates 3 horsepower at TSR = 6 in a 25 mph wind. What is the torque on the powershaft?

Solution: From the equation above,

$$Torque = 2245 \times \frac{12 \times 3}{25 \times 6}$$
$$= 539 \ inch\text{-}pounds.$$

If this is a three-bladed propellor-type rotor, then each blade experiences a bending moment equal to one-third of 539, or 180 inch-pounds. This particular bending moment acts in the plane of rotation—along the thrust direction—and maximum stress occurs at the root of each blade.

Here are a few of them:
1. *Solid wooden blade,* or partially solid carved wooden blade with bolted steel or aluminum hub attachment. Wooden blades can be skinned with fiberglass and resin for improved protection.
2. *Tube spar,* with foam, balsa wood, honeycomb, or other filler, covered with fiberglass and resin. The spar can be made of aluminum, steel, or stainless steel.
3. *Tube spar,* with metal ribs and skin. The metal skin assumes both the airfoil curvature and the blade twist. Rivets and epoxy bonding will keep the skin,

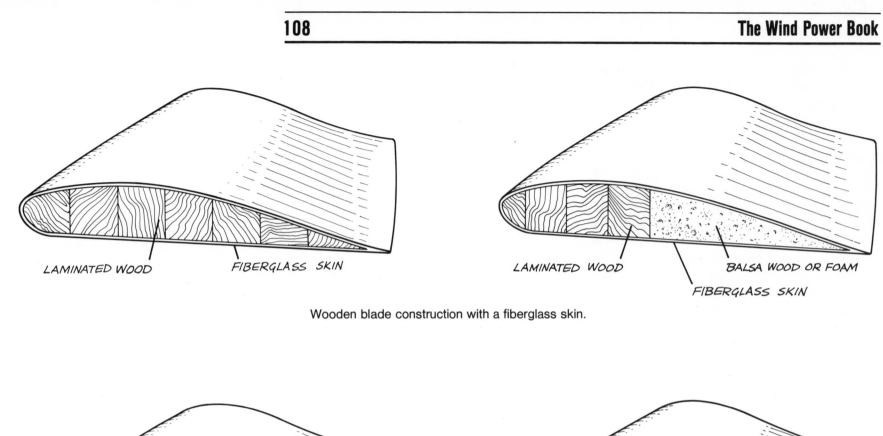

Wooden blade construction with a fiberglass skin.

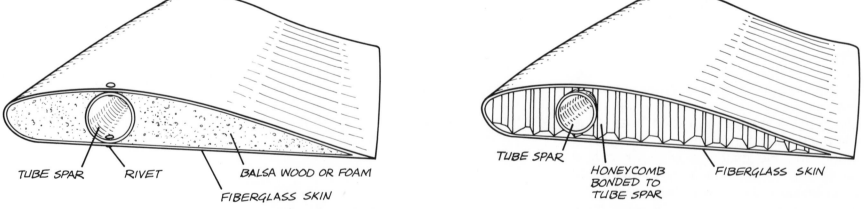

Tube-spar blade construction with a fiberglass skin.

ribs, and spar together. Rivets may be aircraft aluminum, or steel pop-rivets.

4. *Tube spar,* with molded fiberglass skin. The fiberglass skin will be four to eight laminations thick and must be strong enough to avoid flexing in strong winds. A few foam ribs may be bonded inside the fiberglass.

5. *Sail-wing blade.* These blades are made with a tube spar, a stretched-cable trailing edge, and a plasticized fabric membrane (the fabric pores are sealed against air leakage).

The sail-wing membrane changes curvature in response to changing airflow, and can generate high lift very efficiently. The membrane must be stretched fairly tight for best performance. An ideal membrane can be made from the lightweight nylon fabric used for backpacking tents, or the extra-light dacron sailcloth used on hang gliders.

The carved-wood method requires construction skills familiar to most people. Carving wood is easy, fun, and very rewarding. Wood, however, may not be the best material with which to make a windmill blade. Wood is certainly the most readily available, renewable resource, but wood soaks up moisture, and it's difficult to prevent this from happening. If one blade absorbs more water than another, an out-of-balance condition will result. You can see the result of this imbalance by changing the weight of one blade in your calculations for centrifugal force. In the overspeed condition, imbalance can cause the rotor to shake itself to death.

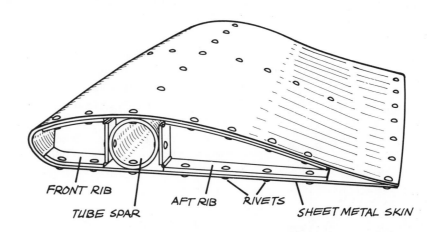

FRONT RIB
TUBE SPAR
AFT RIB
RIVETS
SHEET METAL SKIN

Tube-spar blade construction with a sheet metal skin. Similar construction is used for the skin of modern light airplanes.

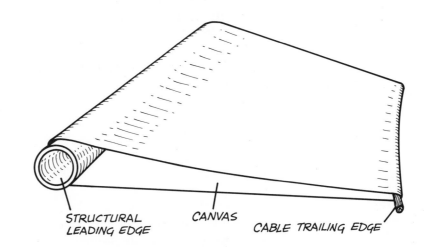

STRUCTURAL LEADING EDGE
CANVAS
CABLE TRAILING EDGE

Typical sail-wing construction. A cable stretches canvas or Dacron fabric from a tube-spar leading edge.

A rivetted aluminum blade used in a straight-bladed Darrieus rotor.

But if you keep the wooden blades sealed, they are a great way to build a rotor.

Lots of people are building windmills with Method 2—a tube spar inside a paper honeycomb filler, skinned with fiberglass. Although honeycomb is expensive, this method is an easy way to build experimental blades. The skills required to work with honeycomb and fiberglass are easily learned, and the results of your efforts will be strong, high-performance blades. The rivetted aluminum structure also yields a blade that is strong, lightweight, and durable. And the skills of rivetting, drilling, metal forming, and bonding are easy to master.

Your first pass at structural design will be based on a guess of the blade weight. Using that estimate, you calculate the loads and size the structure accordingly. With the design that far along, calculate the actual weight of the blade. Chances are, your estimate will need correction, and the blade design will need to be revised. Continue this iterative process until final blade weight equals estimated blade weight.

Flutter and Fatigue

Your wind machine must be able to withstand the various tension, torsion, and bending loads. But how many times can those loads be applied and withdrawn? Continued flexing might result in a crack in the structure and a subsequent failure, called a fatigue failure.

It was a fatigue failure that brought down a blade from the Smith-Putnam wind turbine in 1945. Wartime shortages of materials had forced the machine to remain parked for several years waiting for a replacement bearing. During that time, the parked blades were subjected to vibrations induced by strong winds. These repeated vibrations ultimately fatigued one blade. A small crack developed, which ultimately led to a blade failure.

How does a designer avoid fatigue failures? The secret lies in understanding all the cyclic, oscillatory loads and the response of the materials subjected to them. Most materials have fatigue design curves—called S-N curves—which give the stress level that can be sustained for a given number of cycles. Depending on the type of material used, fatigue-failure conditions often begin when the material is cyclically loaded with more than 20 percent of its maximum allowable stress. Thus, if you expect a tube that will break when loaded to 20,000 pounds to endure cyclic loads, you many not design for more than 4,000 pounds or so of maximum cyclic load. Check the S-N curve for the material being considered to determine where fatigue failure begins.

Flutter can be thought of as a large-scale, catastrophic oscillation. Take a yardstick or plastic ruler and pin it with your palm to the top of the table so that half of its length is sticking over the edge. With the other hand, twang the end of the ruler. It will oscillate and "dampen out" the length of the free end. Make the free end longer, then shorter. The shorter length oscillates much faster than the longer. Notice that longer lengths usually have more deflection, which implies either more bending moment or less stiffness. In this case, the longer the free arm the greater the bending moment. If you change to a different material (e.g., from wood to plastic) less stiffness will cause more bending deflection.

Blade stiffness is crucial to prevention of flutter. When you twang the yardstick, its oscillations dampen out—they converge. In a flutter situation, a windmill blade will not dampen its oscillations. It oscillates more each cycle until the structure fails.

Flutter starts with a bending oscillation (or a torsional, twisting oscillation) that occurs at the *natural frequency* of the blade. You found the natural frequency for each of the different lengths of yardstick by twanging it on the table. You could find the natural frequency of your blade by twanging it, but the blade experiences other forces besides the bending moment. These other forces— especially centrifugal stiffening—complicate matters by altering the natural frequency. Calculating the natural frequency of spinning blades is a complex mathematical problem.

Another difference between the yardstick and a rotor blade in operation is that the wind forces acting on the rotor blade are continually exciting the blade to oscillate. The wind forces and the blade flexing gang up to drive the oscillations beyond the strength of the blade. To prevent such flutter, the blade must be stiff enough that it will not break when its natural frequency is excited by the forces of the wind or by any other cyclic forces.

While the aerodynamic and structural design projects are under way, you must be aware of the various mechanical details that constitute the rest of the wind machine. Apart from structural design of load-carrying beams and powershafts, other features to consider are the governor, yaw controls, and provision for automatic and manual shut-off controls. These features are crucial to the proper operation of your wind machine and to its safety and longevity.

Governor Design

A governor is crucial to the structural life of a wind machine: no governor, no machine. A governor can be a human operator stationed at the machine and trained to take appropriate corrective action in the event of high winds. This was the traditional method of controlling the early grain-grinders and water-pumpers of agricultural societies. The various forms of mechanical governing devices range from simple, spring-loaded widgets to complex, computer-controlled servomechanisms. Here are some of the proven mechanical concepts for governing your wind machine:

1. Aiming the windmill out of the wind. Either turn it sideways or tilt it up—

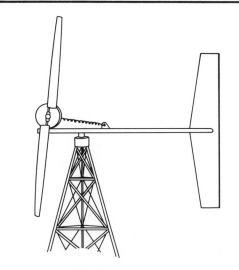

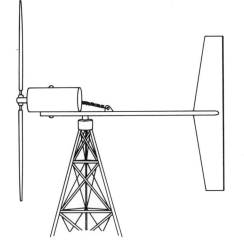

Governing a wind machine by aiming it sideways. This approach is used in some small wind generators and most farm windmills.

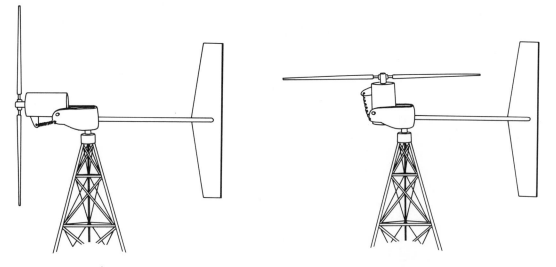

A tilt-up governor. A few modern wind generators use this method.

usually done by allowing the rotor to pivot out of the wind when drag forces are excessive. The American Farm multiblade windmill, for example, is slightly off axis with its tail vane, so that excessively high winds cause it to pivot sideways.

2. Coning, or allowing the blades to form a cone in the downwind direction. This reduces the frontal area, thus reducing the power and rotary speed. The blades can be freely hinged with perhaps small springs to hold them out for starting. Centrifugal force will hold them out during normal rotation, but tempest winds will increase blade drag and cause coning.

3. Aerodynamic control. This method of control has received the most design attention. You can accomplish power control by blade pitch control—rotating the blades to change their angles and reduce power. For blades made with tube spars, you can mount the tubes in bearings at the hub, so that the blade angles can be changed, and provide some means of controlling this angle.

The most common method of controlling blade angle in small installations is to mount a "flyball" on each blade. The flyball tries to swing into the plane of blade rotation against a return spring. This action rotates the blade out of its optimum power configurations and reduces the extracted power. Another approach involves allowing the

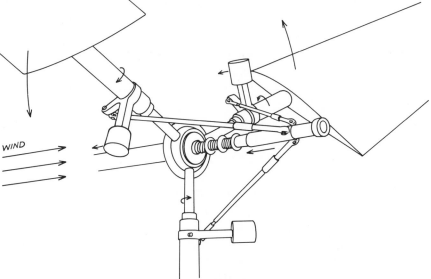

Rotor coning. This and other flexible-hub governors reduce blade loads in high winds.

A flyball governor. At high rotor rpm, the flyballs swing into the plane of rotation, feathering the blades.

WIND

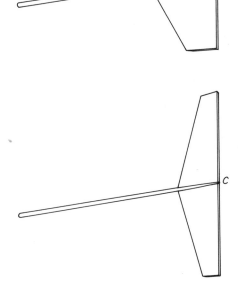

Three different tail-vane designs. Vane C is the most sensitive to shifts in wind direction.

tube spar to slide in and out of the hub. A strong spring holds the blade in against normal centrifugal force, but excessive rotary speed overcomes the spring tension. The blade slides outward but is forced to rotate by means of a spiral groove or slot.

One further method of aerodynamic control uses blade drag to cause the whole blade hub assembly to slide downwind on the power shaft. A spring holds the hub against its stop during normal operation. But when drag is high enough, the hub compresses the spring and slides. The blades are linked to a non-sliding portion of the power shaft so that as the hub slides, the blades are forced to rotate as the linkages extend.

Drag brakes at the blade tips are one more way to harness aerodynamic drag for use in a governor. By mounting small plates there that extend as centrifugal force increases, excess wind power can be dissipated into the airstream—protecting the blades. Spring tension and activation geometry must be carefully adjusted so that the drag brakes come into play only when necessary to limit the rotor rpm.

With any of these methods, the blades and blade-control devices must be linked together so that all blades react together. If one blade's configuration is permitted to vary from that of the other blades during high-speed rotation, severe vibration and balance problems will set in immediately. Usually this is catastrophic. Link all blades and controls together.

Yaw Control

The yaw axis is the directional axis of a wind machine; it is a vertical axis. Vertical-axis wind machines, like the Darrieus and Savonius rotors, need no yaw control. Hence, they are often called panemones—for winds of any direction. For horizontal-axis machines, however, the problem of yaw control is important. The solutions are usually simple. In the windmill shown on page 115, the tail-mounted fin acts like an airplane's rudder to keep the windmill aimed into the wind.

Common types of tail vanes used are illustrated in the diagram. Design A is a bit nostalgic, but it works; design B is a great improvement, and C is the best. The reason is simple. You want the tail vane that is most sensitive and responsive to changes in wind direction. Design C has the highest ratio of vane span (distance from top to bottom of the vane) to chord (distance from leading edge to trailing edge). Such a design is like a glider wing designed to make the most of light updrafts to support the craft aloft without benefit of a motor. Practical ratios of vane span to chord are between two and ten. A typical vane might be five times as tall as it is wide.

Mounting the blades downwind of the directional pivot is another practical method of yaw control. Blade drag keeps the rotor aimed correctly. If you design a machine of this type, balance it to be slightly heavier in front of the pivot by mounting generators or other heavy things up front. Tail-mounted blades are required if you plan to use free-blade coning for gale wind protection. But there is a small problem with tail-mounted blades: the turbulent airflow downwind of the tower causes the blades to vibrate as they pass through this area. This vibration can exacerbate fatigue problems, and should be reduced as much as possible by stream-lining the tower near the blades.

Shut-Off Controls

Shut-off controls allow you to stop the machine for maintenance or in anticipation of destructive gale-force winds. The machine can also be shut off automatically if an abnormal operating condition such as blade imbalance is detected. It's most important to plan some sort of automatic shut-off control into your system. If a blade parts company, this control will keep the rest of the machine from tearing itself away from the tower.

The shut-off control mechanism can be designed into the yaw control (if a tail vane is used) or the governor (if a feathering system is used). A large, reliable brake is used for full shut-off on some systems.

The need for automatic shut-off can result from any abnormal operating condition, such as an unbalanced rotor caused by ice build-up or impact with birds, stones or buckshot. Windspeeds that exceed the design limits should normally cause the governor to feather the blades, lock the

A Dunlite wind generator in Willits, California. The tail-mounted fin keeps this rotor aimed into the wind.

brakes, or tilt the rotor out of the wind. System malfunctions such as generator failure or a blown fuse might create an unsafe condition from which the wind machine should also be protected.

How does an automatic shut-off control work? The mechanical or electrical signal

An Enertech 1800 wind generator installed atop Mount Washington, New Hampshire. Ice build-up is one very good reason for automatic shut-off controls.

that causes a shut-off command or response depends on an abnormal condition being sensed. Blade imbalance or vibrations of any type can be sensed by some sort of accelerometer. One machine uses an iron ball resting on top of a pipe. The ball is tied to a shut-off switch so that, should the machine begin to shake, the ball tumbles from the pipe and actuates the switch as it falls. This vibration sensor might then trip a spring release and swing the tail vane sideways, feather the blades, or apply a brake.

Another, more sophisticated solution is to use an electronic acceleration sensor that sends its signal to an electromechanical device that stops the machine. A wind-speed sensor might be an anemometer or a small vane that, in high winds, overcomes a spring and throws a switch to actuate the rest of the shut-off system. Other system parameters may be similarly sensed, and shut-off effected accordingly.

By now, you may have realized that each part of a wind system should be planned as a whole during, not after, the design process. Each part of the system affects the performance of that system. A poorly planned system will perform below expectations. This chapter has concentrated on the design of machinery sitting atop the tower; the next is devoted to the construction of complete systems.

6
Building a Wind Power System

Details of Energy Conversion, Storage and Delivery

There's much more to a wind power system than just the blades and powershaft. You have to convert the rotary mechanical power into forms more appropriate to the task at hand: the reciprocating (up-and-down) action of a piston well pump, the electrical power in your house wiring, even heat inside a water tank or barn. Some form of transmission takes this power from the wind machine to the actual point of use. And energy storage is required in most systems to take care of the times when there is no wind. Loss of energy and power can occur throughout the system; careful design, construction and maintenance are your only protection against these losses. Finally, some form of backup power is usually required, and controls are needed to make sure that all the parts of a wind system function together properly.

Wind systems have become more complex. Skills required to build wind machines have grown, and designs have become more sophisticated. Flat airfoils evolved into aerodynamically curved shapes; bicycle chain drives were replaced by gear boxes; wooden pushrods became electrical wires. The many practical wind systems already developed provide us a firm basis for further work. With new technology and fresh thinking, we can design many new systems to meet current and future energy needs.

The basic elements of a wind power system are illustrated in the accompanying diagram. The blocks represent system components; the arrows indicate the transmission of power or energy. All systems, even the simplest, have a power source and a user. The power source may be an old multiblade windmill, and the user a nearby sprinkler or an irrigation ditch. Rotary shaft power is converted to reciprocating mechanical power to drive a piston pump that lifts water out of the ground to the irrigation ditch. Such a system is not very different from the earliest wind power applications.

But suppose you wish to irrigate during the night, and the wind blows mostly during the day. Or, suppose you need to irrigate the crops every day, but the wind blows only three days each week. In each case, you need to store the pumped water until it is needed. This pumped water is a form of energy storage, just like the chemical energy stored in fully charged batteries. Most applications of wind power require some form of storage so that power is available when the wind has died.

Thus, most wind power systems include a wind energy converter (the wind machine), a user (your application), and some form of energy storage.

A wind power system is often planned for applications that require a constantly available source of energy. Suppose, for example, that wind is used to power a remote data sensor and transmitter. Clearly, certain applications for remote data collection, such as forest fire prediction, require an uninterruptable energy source. Systems that must always provide power, regardless of wind conditions, must be provided with a backup energy source. For the remote transmitter,

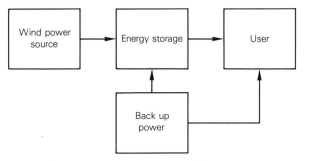

Block diagram of a wind power system. A complete system includes energy storage and a provision for generating back-up power.

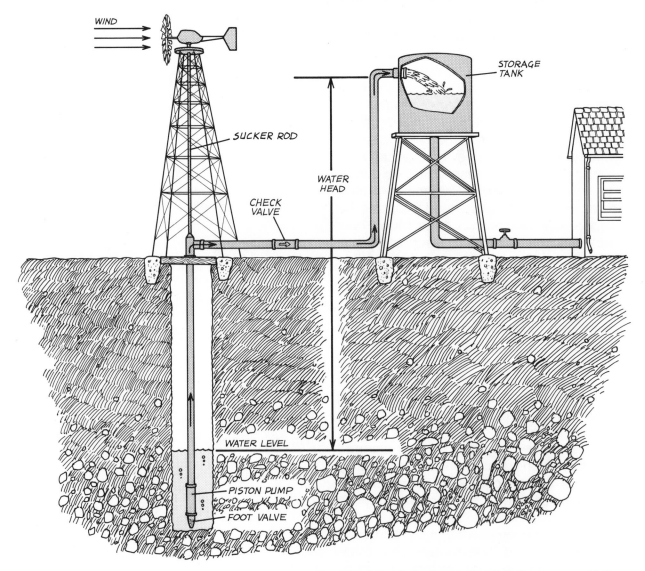

WIND

SUCKER ROD

STORAGE TANK

CHECK VALVE

WATER HEAD

WATER LEVEL

PISTON PUMP

FOOT VALVE

A complete water-pumping system. Wind power lifts water through the "water head" and overcomes friction losses in the pipes.

backup might be a propane-powered generator; a switch that, when thrown, connects the user to utility mains; or jumper cables to the electrical system of a handy automobile.

In any case, you must pay attention to the controls designed or built into the system. You cannot attend the wind machine's needs at every moment, so it's important to consider its ability to shut itself off when high winds occur, the batteries are fully charged, or just before the water storage tank overflows.

Electronic voltage regulators like those used with automotive alternators will prevent overcharging of batteries by reducing field current to the alternator, thus reducing the alternator's output and its load on the power-shaft. The batteries won't be overcharged, but the load on the wind rotor will be reduced, thereby increasing the importance of an adequate governor to protect the rotor under all conditions. Such governors can be designed to use (1) mechanical forces generated by springs and flyweights or (2) electric or hydraulic servomechanisms controlled from electronic or other sensors. Or, as in the case of the farm water-pumper, wind forces acting on the rotor tilt the rotor sideways out of high winds.

The final design and selection of components that make up the blocks in a diagram of your system depend mostly on the application you select for your wind system. Water-pumpers have different design problems and solutions than battery-charging systems. If water is to be pumped with an electric well pump, you must consider the problems of both water pumping and electricity generation. Let's look at some of the available solutions.

Water-Pumping Systems

Many kinds of wind machines have been built to pump water, including the Cretan sail-wing, the Dutch four-arm, and the Halliday machine. But perhaps the most familiar wind-powered water-pumper is the multibladed farm pumper that is a common sight in rural America and parts of Australia and Argentina.

Rotary powershaft motion is transmitted, sometimes through a speed-reducer gearbox, to a crankshaft that converts it into the linear reciprocating (up-down) motion of the sucker rod, a vertical shaft extending down the tower to the pump below. Sucker rods got their name from the fact that many deep-well piston pumps lifted the water only on the upward, or sucking, motion of the rod. Such pumps are called single-acting piston pumps; double-acting pumps lift water on both the upward and downward strokes. Lifting on the upward stroke places tension loads on the sucker rod, while pumping on the downward stroke applies compressive loads that can easily buckle a long, thin rod—especially when a strong wind is busy bending the rod ever so slightly. According to many farmers, most of the maintenance required by water pumpers with double-acting pumps consists of replacing sucker

rods and pump seals.

A foot valve at the bottom of the well serves as a one-way inlet valve. Usually, a one-way check valve is installed between the well casing at the pump outlet and the tank. Installing a check valve here reduces water loss backward through leaking pump seals.

Among the various other pumping schemes are paddles that splash water over the edge of shallow wells, disks drawn by a rope through a tube, and various screws and centrifugal "water slingers." Such low-technology pumps have been built by Dutch millwrights for draining land behind dikes, and by Asian farmers irrigating rice paddies and gathering salt from seawater on large solar-heated salt ponds. These machines are appropriate solutions for many of today's energy needs, particularly in remote, rural areas.

A recent method of wind-powered water pumping uses compressed air to transfer the power to the point of use. Rather than a sucker rod descending the tower, compressed air is pushed through a tube and used to raise water out of the ground. A windmill drives a conventional compressor like one on a spray-paint compressor. As the rotor spins, air is compressed and sent in hoses to a storage tank. This compressed air drives then a submerged piston, bladder, or rotary pump. Such pumps are either adaptable to or specifically designed for this purpose. Air-driven piston pumps tend to be elaborate devices. The air must first

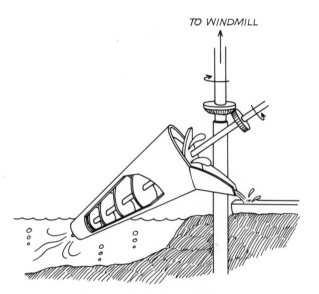

TO WINDMILL

The spiral-screw water pump has been used in Dutch machines to drain large land areas.

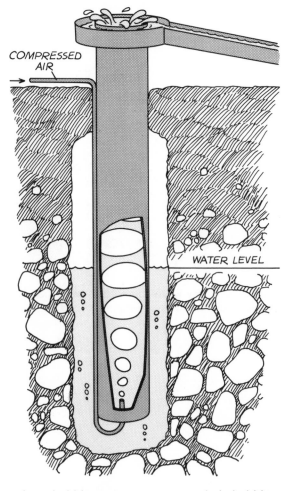

In a bubble pump, compressed-air bubbles expand to the pump diameter as they rise, forcing water to the surface.

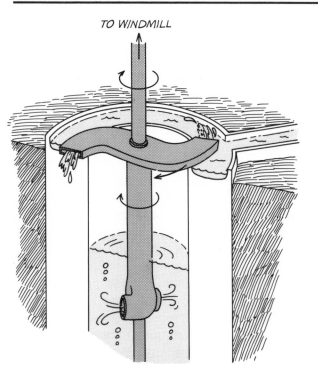

A centrifugal pump can lift water from a shallow well or a pond.

drive one way and then the other, each time being valved on both the intake and release (exhaust) sides.

A much simpler pump is the bubble pump. You can visit a tropical fish store and see one of these at work. Just blow bubbles into a pipe and they expand as they rise, pushing water ahead of them. They must be fully expanded to the walls of the pipe before they reach the water surface level, or they will "burp" and fail to pump water out of the ground. With bubble pumps, you can raise

water about twice the height the pipe extends below the water surface.

Unfortunately, bubble pumps operate at terribly low efficiencies of 20–50 percent. They waste up to 80 percent of the compressed air just to blow big enough bubbles to pump water and push that water against pipe friction. But there is a lot of room for design improvements, and bubble pumps are virtually maintenance-free. The compressor driven by the windmill is not maintenance-free, so it should be installed in a place conducive to easy maintenance. Bubble pumps are *not* good at transporting water horizontally. Burping wastes so much power that horizontal pumping is all but impossible. You can, however, pump water high enough vertically to allow gravity to carry the water along a trough to the tank or user.

Water Storage

Generally, pumped water must be stored for a while (unless it is being used directly for irrigation) because the wind does not always blow when you need the commodity it helps deliver. Early American pumped-water storage schemes included ponds for cattle watering, stock-watering tanks made of wood or metal, and redwood water tanks with a capacity of 500 to 10,000 gallons. The redwood tanks were built near or into the house for which the water supply was intended. About the only changes that have come along are galvanized metal tanks, fiberglass tanks, and plastic tanks.

Two storage considerations are important: maintenance of a given quantity of water, and maintenance of enough water pressure to allow the water to be used as intended. Stock-watering from ponds requires only a pond full of water—no pressure. But a shower works best if the water tank is at least 5 feet above the shower head. Some water pressure can be maintained by charging a pressure tank, as most electric well pumps do. Often a flexible bladder expands inside the tank to accommodate the extra volume of water. This arrangement allows you to have plenty of pressure—for a short time—when you turn on the faucet. A 50-gallon accumulator tank might be sufficient for all your water storage needs, but most people need a larger tank.

From an architectural viewpoint, large water tanks are less desirable than 50-gallon demand accumulators. But in the days when wind power was the prime source of pumped water, tanks as large as 10,000 gallons were a common architectural form, often integrated directly into the dwelling. Nowadays the trend is definitely toward the smaller tanks.

Designing Water-Pumping Systems

In order to size your water pumping system you will have to estimate your power requirements. How much water do you need? How fast, when, and how often do you need it? By answering these questions, you arrive at an energy budget expressed in gallons of

A water-pumping windmill of early California. The storage tank was built into the structure itself.

Power for Water Pumping

The power required to pump water is proportional to the flow rate and to the pressure against which the pump is working. This pressure is usually expressed in terms of the "head," which has two contributions: (1) the height that the water must be pumped from groundwater level, and (2) an extra contribution, called the "friction head," due to friction in the pipes retarding the flow of water. A fairly accurate value for the power required can be calculated using the formula:

$$Power = 0.00025 \times \frac{G}{E_p} \times (WH + FH),$$

where

- G = water flow rate in gallons per minute;
- E_p = mechanical efficiency of the water pump;
- WH = water head—the vertical height in feet from groundwater level to tank inlet; and
- FH = friction head in feet.

This formula gives you the power in units of horsepower; to get the answer in watts, multiply by 746.

This equation is meant to work with the average flow rate and to give you the average power required by the pump. One simple way to establish an average flow rate is to estimate your need for water, expressed in gallons, and divide this need by the number of hours

you expect the wind to produce usable power during the same time period. Do this on a daily, weekly or monthly basis—depending on the results of your wind survey. To get the flow rate in gallons per minute, then, use the formula:

$$G = \frac{Gallons\ Needed}{60 \times hours\ of\ wind}.$$

Friction losses depend on the pipe length L (in feet), the pipe diameter D (in inches), the number N of pipe joints and corners, and the flow rate G. The formula for friction head FH is:

$$FH = \frac{L \times G^2}{1000 \times D^5} + 2.3 \times N.$$

The pipe length L includes all pipes down in the well, across the pasture, and up the hill or into the tank. If pipe diameter changes along the way, as it usually does, this formula must be used separately for each different length of pipe, and the results added together to get the total friction head.

Example: Suppose the depth of water in a well is 100 feet below the ground level. A 3-inch pipe brings this water to the surface. The water passes through one elbow joint and into a 1-inch diameter pipe running 200 feet along the ground, then through a second elbow joint and up 14 feet before

passing through a third elbow and into the tank. Suppose the farmer needs 2700 gallons of water per day and there are an average of 6 hours of usable wind per day. How many watts of wind power must be supplied to a water pump with 75 percent efficiency?

Solution: The average flow rate needed is

$$G = \frac{2700}{60 \times 6} = 7.5\ gallons/min.$$

The friction head in the 3-inch pipe is

$$FH = \frac{100 \times 7.5^2}{1000 \times 3^5} + (2.3 \times 1)$$

$$= 2.3\ feet.$$

The friction head in the 1-inch pipe is

$$FH = \frac{214 \times 7.5^2}{1000 \times 1^5} + (2.3 \times 2)$$

$$= 16.6\ feet.$$

The total friction head is the sum of these two contributions or $FH = 18.9$ feet. Then the pump power required is

$$Power = 0.00025 \times \frac{7.5}{0.75} \times (114 + 18.9)$$

$$= 0.33\ hp.$$

To get the answer, you just multiply by 746, so the power requirement is 246 watts.

water, which you can convert to foot-pounds or kilowatt-hours. Usually there is plenty of room for water conservation. Do you really expect to take half-hour hot showers, knowing there is no wind blowing?

Careful study of your planned water usage may reveal important conservation areas. Drip irrigation combined with soil moisture sensors can tremendously reduce the perceived water requirements of many crops. As opposed to large stock troughs, the automatic stock-watering devices commonly available at feed stores reduce evaporative losses and algae growth. Emulsifier showerheads, water softeners, and other devices can readily reduce domestic water consumption. Conservation should be part of any equation used to calculate an energy budget.

Next you will need to determine the height your water must be raised and the flow rate required. The power required to pump the water is proportional to the flow rate and the water head—the height above groundwater level to which the water must be pumped. The lower the flow rate or water head, the less power required.

How do you measure groundwater level? A string with a weight and float attached will do, if the well driller isn't around to give you his findings. Note, however, that well depth probably will change with season. Through careful prospecting, you might find a shallow well site; with any luck, such a well site will be near the intended water use.

Long pipe runs take their toll of power through friction. Friction increases with length, with the square of flow rate (twice the flow rate means four times the friction). It decreases with the *fifth* power of the pipe's diameter (cutting a pipe diameter in half increases the friction 32 times). Take a 50-foot garden hose and measure its flow rate with a faucet turned on full. Add another 50-foot length and measure the new flow rate. Add enough hose, and the flow will be reduced to a trickle. Then try this test with a smaller diameter hose and see what happens.

To reduce pipe friction, select a pump that produces low flow rate, and a pipe with the largest possible diameter and the fewest joints. You can't always choose a pump on the basis of flow rate, however. For a pump to absorb any amount of power from the windmill and translate this power into water flow, it will do so at high pressure and low flow rate (deep well) or low pressure and high flow rate (shallow well), or some level in between. This happens because the power required to lift water is a function of well depth, desired flow rate, and friction. Increase any of these and you increase the power required. If power is limited and water depth increases as the summer ends, the flow rate must decrease to compensate. Jet pumps, which use a jet of high-pressure water to lift well water to the pump where it can be sent along its way, are sold on a basis of horsepower and well depth. You can mix-and-match jets and pump motors if you wish, but if you deviate much from recommended settings, you will be ignoring a

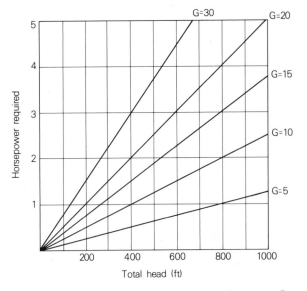

Pump power needed to lift water at a flow rate G, in gallons per minute. To get the wind power needed to drive the pump, divide by the pump efficiency—about 70 percent.

tremendous amount of research that has gone into improving the efficiency of these units. Your best option to reduce flow rate is to use the largest pipe size possible.

Tuning a water pump system often reveals interesting characteristics. Take the case of a piston-type water pump, which operates off a crankshaft in an oscillatory fashion. The speed at which the piston moves along in the cylinder changes constantly—from zero at the instant it changes direction to a maximum speed at the midpoint of its travel. It's this maximum flow rate that needs close examination. For an average flow rate to equal your calculated values, the maximum instantaneous flow rate must be much faster—fast enough, in fact, that fluid friction really becomes important. The pump will not operate efficiently unless the pipes can handle this surge in flow rate.

Other than using large-diameter pipes, there is only one solution. Adding a *surge chamber* near the pump will solve the problem at low cost. What is a surge chamber? It's a large tank of trapped air plumbed to the water line. Whenever a flow surge puts pressure on the long water line, most of that pressure instead compresses the air trapped in this chamber. Water enters the surge chamber quickly but leaves it slowly, under pressure from the air it compressed, during the time between surges when the piston is pumping more water into the chamber at a much lower instantaneous flow rate.

A low-cost surge chamber can be made from recycled refrigeration charging tanks discarded by refrigeration repair shops. But a small problem remains: aeration of the water in the chamber. Water under pressure will, over a period of time, absorb the trapped air, and the tank will fill up with water. No surge capacity will result unless the tank is periodically burped—approximately the opposite of burping a baby. There, you coax air out; here, you let air in. A commercial surge tank comes equipped with a rubber bladder that solves the problems by separating air from water.

By selecting large pipes and reducing the number of joints in the pipes—especially corners—you minimize system losses. Minimizing seal leaks, eliminating evaporative losses, and selecting efficient pumps are also effective means in optimizing a system. Maximizing energy availability in the design process (i.e., matching wind machine to wind availability) should then yield the energy desired. Also, you should select an aerodynamic design consistent with the level of technology you wish to employ. All of this is followed by construction, installation, testing and tuning of the system.

Wind-Electric Systems

Because electricity is a low-entropy, high-quality form of energy, it can be readily adapted to many end uses. Hence, there are many different systems that can use a windcharger as the power source. These include charging batteries for domestic power and lighting, driving an electric water-pump, powering a remote electronic sensing device, and powering wind furnaces for home and farm heating. The possibilities are virtually limitless.

The earliest domestic windcharger powered a radio set and sometimes a light. Batteries were used for energy storage, and either a gasoline or kerosene generator was also on hand to boost the batteries. Or the farmer carted them to town occasionally if the wind didn't fully support the electric demand. In town, an appliance and implement repair shop—usually also a windcharger dealer—would recharge the batteries while the farmer ran his errands. I've often recharged my batteries in the trunk of my car while enroute to town.

The electric system used by many farmers prior to rural electrification was a direct current (DC) system, like the one illustrated here. A two-to-four-bladed windmill turned a generator, often not too different from an automobile generator, that produced electric current flowing in a single direction along wires connecting it to the battery bank. Today, the same DC system is used in most domestic wind-electric systems. This system has no provision for alternating current (AC) loads, such as refrigerators, washing machines, and the like. The old radio sets used a mechanical vibrator inverter in their internal circuitry to convert the usual 32 volts DC to an AC voltage that the radio power supply could use.

With the invention of transistors and integrated circuits, the need for vibrator inverters

in radios has been eliminated, but the use of inverters in general has not. Today, wind-electric systems are being used to power a greater variety of loads than did the older windchargers. Historically, direct-current devices such as lights, radios, electric shavers, and electric irons were the usual loads. Now microwave ovens, color TV sets, freezers, refrigerators, and washers are becoming common. Electrically, these 110-volt AC loads are very different from their 32-volt DC ancestors. Because of this difference, the selection of an appropriate wind system design is more complex than it was in the past.

Optimizing a wind-electric system design, or planning that system for maximum energy production at least cost, involves selecting the smallest component or subsystem that will do the job safely. A wind-electric system is completely optimized when no energy storage is needed—an unlikely situation. Energy storage—traditionally batteries—wastes an enormous amount of energy and often runs system cost beyond practical limits. Arriving at a site and system design that closely couples load timing with wind availability would be ideal. Often as not, though, you need energy during the evening, and the winds occur in the morning.

Generators and Transmissions

Wind power is converted to rotary shaft power and finally into electrical power by a generator or alternator. The generator might

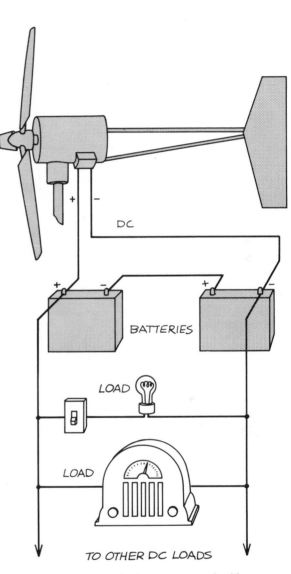

Direct-current electrical system used with many windchargers. The batteries are wired in series.

be AC or DC but an alternator generally produces AC. Most automotive alternators contain AC-to-DC rectifiers that supply DC current to the automotive electrical system. AC power can be generated directly at the wind generator, or AC can be made from DC using an inverter.

Alternating current comes directly from wind machines that are designed with either induction generators or synchronous generators. Each is really a motor that runs at an rpm that is governed by the load and the 60-cycle line frequency. Usually, these motors run at 1,750 to 1,800 rpm, with the higher rpm occurring when the motor is fully unloaded.

Direct current is generated by either a DC generator or an AC alternator with rectifiers. Traction motors used in golf carts, fork lifts, and electric cars are suitable for use in a wind generator. Usually these motors have commutating brushes that carry the full output current of the generator. Alternators come in a variety of sizes and types, from small automotive types to hefty industrial alternators. Automotive alternators are not very efficient (about 60 percent compared with over 80 percent for the industrial variety and DC traction motors) and must be driven at high rpm. They are cheap, however, and find their way into lots of wind projects.

The alternator or generator is coupled to the rotor through some form of transmission that serves to speed up the relatively slow-turning powershaft to the higher rpm required by the generator. Some machines have been built—the old Jacobs, for example—with a large, heavy, slow-turning generator that could be coupled directly to the rotor. Some newer alternators are being tested that can also couple directly, but most generators and alternators spin so fast that a speed-up gear ratio is needed.

What form does the speed-up mechanism take? Chains, belts, and gearboxes are the devices usually employed for the speed-up task. Gearboxes that are readily available at local tractor-bearing, chain, and pulley stores are suitable. These gearboxes start out as speed reducers, but windmill designs simply run them backwards as speed increasers. When running a gearbox backwards, the horsepower rating—which would normally equal the maximum rotor horsepower—should be conservative: select a gearbox larger than needed.

Chain drives may be cheaper than gearboxes, but oiling and tensioning requirements can make them a less desirable solution. Some designers have successfully enclosed their chains and sprockets in sealed housings with splash oil lube; a spring-loaded tensioner coupled with the oil bath could make a chain drive a reasonable part of your project.

Several belt drives are available; the toothed belt and the V-belt are the most common. Toothed belts and pulleys are nearly as expensive as gearboxes, and these belts can be temperamental. If their pulleys are not properly aligned, the belts slide off. If the torque is very great at low rpm, they hop teeth and eventually self-destruct. Cold weather can destroy them too. V-belts tend toward high friction, which means an inefficient power transmission, but they are by far the cheapest solution. Still, gearboxes are my favorite, most easily obtained transmission.

Storage Devices

Since you are likely to need energy during periods of no wind, some form of energy storage will be required. The three main types of energy storage available to you are thermal, mechanical, and chemical. With thermal storage, you simply convert wind-generated energy into heat by using an electric-resistance heater probe or, more directly, by stirring a tub of water. You then store this heat as hot water, a warm bed of rock or gravel, or molten heat-storage salt. These techniques are discussed later under the subject of wind furnaces.

Mechanical energy storage involves coaxing a heavy object into motion or lifting it so gravity will later return it. A favorite analogy of mine is to think of wind-generated electricity powering an electric motor mounted at the top of a 100-foot pole and busy lifting, by cable and winch, a 1955 Oldsmobile. When the car reaches the top of the pole, the energy storage device is said to be fully "charged." If juice is needed, the Olds is allowed to drop—spinning the motor in reverse as it falls. The motor becomes a generator, providing electricity on demand.

When the Olds hits the ground, the storage cell is said to be "dead." Pumped water is a better way to store energy that I will discuss later.

A more familiar energy storage cell is the flywheel. Instead of lifting a bit of American history 100 feet above everybody's heads, a spinning disk stores energy. Conventional wisdom says that this spinning wheel should be large and quite heavy. The more weight, and the faster it spins, the more energy the heavy wheel can store. More recent thinking has resulted in the development of super flywheels. These flywheels are not very heavy, but they spin incredibly fast—30,000 rpm or more, compared with about 300 rpm for the heavy wheel. Because energy stored in a flywheel is directly proportional to weight (double energy for double weight) but increases with the *square* of the rpm (quadruple energy for double rpm), an enormous amount of energy can be stored in a really fast-spinning wheel—but not without penalties. At such high speeds, air friction is considerable, so super flywheels are typically installed inside a vacuum chamber. Also, the bearings must be very precise devices, carefully designed and built. These problems notwithstanding, super flywheels may eventually compete with batteries for storage of wind-generated energy.

There are two approaches to chemical energy storage. One takes electrical power and splits a compound, say, water, into its constituent parts—hydrogen and oxygen. These constituents are stored separately

Commonly used to power golf carts, this rechargeable battery can be used for storage of wind-generated electricty.

Battery banks are the most common energy storage for wind-electric systems.

and later recombined to produce electrical power as needed in a fuel cell. These cells are becoming more available recently and, while expensive, offer a means of using the energy of chemical bonds to provide the needed energy storage. Another approach to chemical energy storage is the traditional battery storage device. The rechargeable battery is the only type being considered for wind energy storage. Metal plates inside these batteries act as receptors for the metal atoms that plate out of the electrolyte (the acid in lead-acid batteries, for example) as the batteries are being charged. When the battery subsequently discharges, these

metals return to solution in the electrolyte—releasing electrons and generating direct current at the battery poles.

There are dozens of different batteries—each named by the type of plates and/or electrolyte used. The most common are the lead-acid and the nickel-cadmium batteries. Lead-acid batteries are used in cars, golf carts, and other common applications. Nickel-cadmium, or nicad batteries as they are called, are used when higher cost is not too great a penalty and when lower weight and improved tolerance to overcharging are required. Airlines use nicads.

You can use either type in your wind system, but you must be careful not to overcharge a lead-acid battery, or discharge it too fast. The nicad battery can stand these abuses, but it has one quirk worth mentioning—a memory. If you discharge a nicad only half-way each time, eventually half-way will be as far as it will go. If your electric shaver uses nicads, it's best to run it dead periodically. Whatever the battery used, you can reduce storage losses by using clean terminals and well-maintained batteries. You should also try to have sensible discharge rates, because high discharge rates lower the storage efficiency.

Inverters

Electronic, stand-alone inverters convert direct current into alternating current at a voltage and frequency determined solely by each inverter's circuits. Inverters that turn, say, 12 volts DC into 110 volts AC are available in a variety of amperage ranges and voltage and frequency accuracies. For example, a really cheap 4-ampere inverter—that's 110 volts at 4 amps, or about 400 watts—is available in the recreation vehicle market for under $100 (1980). Voltage output may vary from 105 to 115 volts, and frequency from 55 to 65 cycles per second. And the output is not the smoothly oscillating sine wave of expensive inverters, but rather a square wave, typical of cheap inverters. Square-wave AC is useful for operating motors, lights, and other nonsensitive devices, but stereos and television sets tend to sound fuzzy because of the effect square waves have on their power supply.

More expensive inverters are available with close voltage regulation, a quartz crystal to control frequency, and a sine wave output. For applications where the AC output will power electronic devices, this type of inverter may be necessary.

Where voltage and frequency are not important, but sine wave output is, a motor-generator type of inverter can be used. Here, DC powers an electric motor that in turn spins an AC generator. AC generators produce sine waves, but heavier loads slow the motor down and cause frequency and voltage drops similar to those of the cheap electronic inverters. Inverters tend to operate at highest efficiency at or near their rated power. Because your AC loads aren't always on, or on at the same time, it might be

possible to enhance overall system efficiency by using small inverters at each important load. This is the approach used by Jim Cullen in his solar- and wind-powered home in Laytonville, California (See Chapter 2). These local inverters should be selected to operate only at their peak efficiency. Be sure to pay attention to their surge ratings. Most loads, especially motors, draw a surge of current several times their rated amperage for a few seconds during starting. This current surge can destroy an inverter that's not designed for it. Consult Jim Cullen's book (see Bibliography) for more details on the installation and use of stand-alone inverters in a house wired for direct current.

Synchronous inverters perform a slightly different task than the stand-alone types. They still turn DC into AC, but they drive an existing AC line. The AC output from a synchronous inverter is fed directly into an AC line such as your household wiring. AC current is available from an existing source, so this inverter must synchronize its output with the AC line and operate in parallel with the other source. To be fully synchronized, the wave forms—sine waves in the case of utility power fed into your household wiring—must match. In fact, most synchronous inverters use the wave form they are working with to establish internal voltage and frequency regulation.

To get AC into your house from a wind generator, then, you either start with AC at the windmill, or turn DC into AC with an inverter. Either way, a major safety aspect

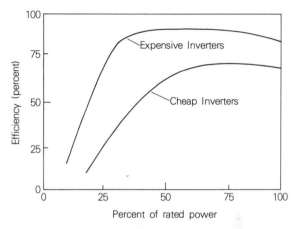

Efficiencies of two different inverters. A well-matched inverter operates at more than half its rated power level.

The Gemini synchronous inverter. This electronic device converts direct current into AC power synchronized with utility power.

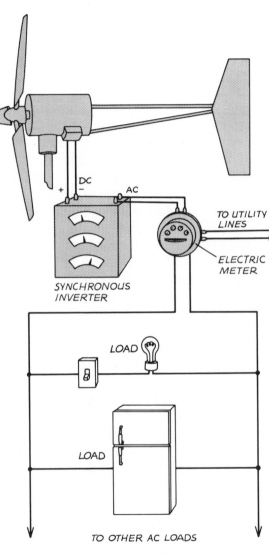

A synchronous inverter couples a DC wind generator to both the utility grid and the house wiring system.

must be considered: if you shut house power off to service the wiring, the wind-generated AC source must be shut off, too. Failure to do so can be fatal. In gaining approval from your public utility to link an AC wind system with your house, the safety issue will be the primary topic of discussion.

Designing Wind-Electric Systems

By now, you're probably considering some sort of end use, or goal, for wind generated electricity, such as powering your house, heating a barn, or whatever. What sort of wind-electric system plan should be used to reach that goal?

The first choice the designer of a wind-electric system must make is whether he or she wants an AC or a DC system. Then, a suitable plan takes account of several important factors:

- The nature and timing of the load
- Rated power of the windmill
- Constraints of cost and availability.

Most important is the nature and timing of the load. Look at a typical domestic load plan. A small house with a few inhabitants has the usual array of consumer gadgets. The energy and power requirements for this household are summarized in the table.

Your needs for electrical energy and power depend entirely upon the appliances you have and how you use them. To determine these needs, you must examine all your appliances and monitor or estimate their use patterns. Whenever possible, you should determine the power drawn by each appliance—its rated power. This will usually be written, in watts or horsepower, on a label somewhere on the appliance. If the appliance is not labeled, an AC wattmeter might be used, or you could consult the manufacturer to determine the rated power. Appendix 3.5 contains extensive tables of the energy and power requirements of many electrical devices.

If low-cost system design is your goal, you will have to have some estimate of the timing of the loads placed on your system. If all the devices listed in the example illustrated here were used at the same time, about 3,500 watts of electrical power would be required. In a 110-volt AC system hooked up to the utility lines, that's only about 32 amps (watts = volts × amps). But in a 12-volt wind-electric system, you'd need to supply almost 300 amps! In addition, the motors in appliances like refrigerators and freezers draw up to five times their rated power for a few seconds after starting. This surge doesn't contribute much to the overall energy needs, but it might well put an excessive load on batteries, inverters or other system components.

The methods you can use to estimate load timing all start with an inventory of your electrical devices as illustrated here. Consult Appendix 3.5 for more detail about exact procedures. Once the monthly energy use has been estimated (about 300 kWh in our example), compare it to your monthly electric bill. If they don't compare, something

Electrical appliances in a small household. The energy and power needs of these devices are summarized in the table below.

is wrong with your estimate, and you should try again.

Virtually all of the devices commonly available off-the-shelf at consumer goods stores will require 110-volt alternating current. The on-line AC frequency should be 60 cycles per second with only a slight variation in frequency allowed. If the house is so equipped, you should use one of the AC load systems shown on the next page.

ENERGY AND POWER REQUIREMENTS FOR A TYPICAL HOUSEHOLD			
Appliance	Hours Used (hrs/month)	Rated Power (watts)	Energy Use (kWh/month)
Refrigerator	500	360	180
Kitchen light	120	100	12
Bedroom light	100	100	10
Porch light	100	40	4
Living room light	120	100	12
Bathroom light	100	75	8
Television	120	350	42
Microwave oven	15	1500	22
Slow cooker	40	75	3
Misc. kitchen devices	8	250	2
Blow dryer	8	500	4
Misc. bathroom devices	20	50	1
TOTALS		3500	300

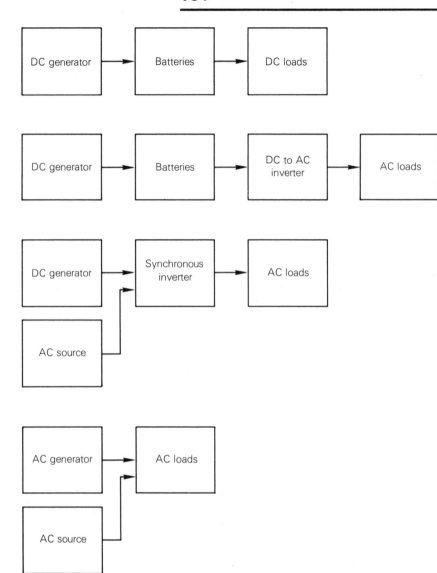

Block diagrams of four basic wind-electric systems.

If your load is off-line, or not connected to a utility grid power line, then you should also select one of the AC systems but 60-cycle power is not required. Generally, DC energy storage and an inverter are used with off-line loads.

If the household load is on-line, then the source of backup power could well be the utility lines. By synchronizing frequency of the wind-generated electricity with the line frequency, the wind system can operate in parallel with the utility power supply. The power required by the domestic loads can be supplied entirely by wind, if available, or by the utility if not. The two power sources could even operate in tandem.

Remote electronics applications are often ideally suited to wind power. Telecom-munications equipment, environmental mon-itoring sensors, and forest fire spotting equipment are a few examples of installations requiring electrical power that is usually unavailable from utility lines. These loads are almost always off-line and require a DC wind system.

A principal difference between remote electronics wind systems and similar systems used for domestic purposes is the nature of the load. A windmill designed for remote work must be highly reliable, with controls to make it self-sufficient for extended periods. One might reasonably expect a domestic wind system owner to shut down his gener-ator manually in a storm, but sending some-one backpacking into the woods in the face of a storm is out of the question. Fully automatic controls can be built into either system, but cost constraints might rule out their use in most domestic systems.

Wiring diagrams for direct-current systems range from simple to complex. Direct current from the wind generator charges the two batteries shown in the diagram on page 127. There the batteries are wired in series—the positive terminal of the first battery is linked to the negative terminal of the second. If each battery is rated at 6 volts, the two together will generate the same voltage as a single 12-volt battery. By linking them in parallel, as shown in the first figure at right, the voltage output is limited to 6 volts. If more batteries are needed to store more energy at the higher voltage, they would be linked in series-parallel to each other, as illustrated in the second figure here.

If a good timing match exists between need and resource, most of the wind-generated electricity will directly power the load. If there is no need, the batteries will be charged, and some energy will be wasted due to the inefficiency of charging the batteries.

An electronic circuit that monitors system voltage may be needed to prevent wasting wind-generated electricity when the batteries are fully charged. Suppose the battery is fully charged and wind power is available but no electricity is being used. The load monitor senses this condition because, when the battery is fully charged, voltage in the charging circuit will tend to rise above normal charging voltage—a 12-volt battery

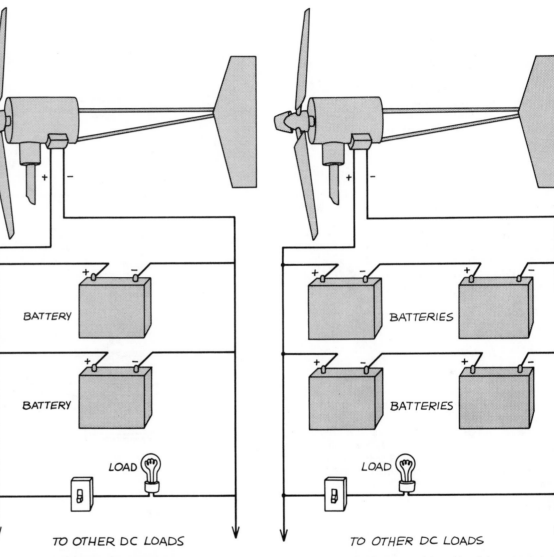

Parallel battery wiring increases the ampere-hours you can store but lowers the voltage.

Series-parallel battery wiring. Use this approach to store more electrical charge at higher voltages.

Sizing a Heater Probe

To determine an appropriate value of the electrical resistance needed in a water heater probe, use Ohm's Law. This is a tried and proven formula that requires only a knowledge of the current (in amperes) you expect to flow through the probe and the voltage of the system:

$$Resistance = \frac{Voltage}{Amperes}.$$

The current passing through the probe is just the excess current from your wind generator that you want to dissipate.

Example: *Suppose your wind system has a 12-volt generator, with a maximum charging voltage of 14.4 volts, that is rated at 80 amperes. What resistance do you need in the heater probe to dissipate half this rated current?*

Solution: *Using Ohm's Law with the maximum charging voltage,*

$$Resistance = \frac{14.4 \; volts}{40 \; amps}$$
$$= 0.36 \; ohms.$$

You should then purchase a heater probe with a resistance of 0.36 ohms. If your system instead could generate 110 volts at a charging current of 20 amperes, and you wanted to dissipate the full current in a water heater, the probe resistance would have to be 5.5 ohms. Note that the 0.36 ohms in the previous example would look like a short circuit to this 110-volt system.

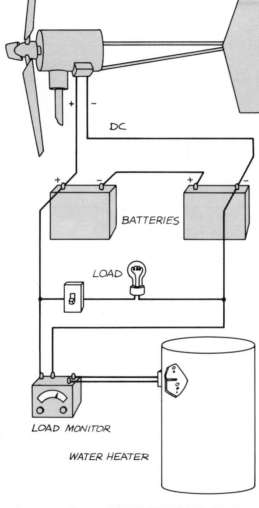

Series battery wiring with a load dumper. When the load monitor senses an overcharged battery, it diverts current to an electric water heater.

will charge up to about 14.4 volts. At this voltage and above, overcharging occurs, which can damage the battery. As long as the charging voltage remains below a set voltage, say 14.3 volts, the monitor continues to charge the batteries. Once the voltage rises above this point, the monitor switches on a relay that adds a new load to the circuit. This load gives the excess wind power a place to do some work without overcharging the batteries, thus avoiding any waste or battery damage. After the new load has been applied and charging voltage has dropped, the load monitor will switch the relay to off and begin charging the batteries again.

What sort of extra load am I talking about? It can be an electric resistance heater or a backup heater probe in a solar storage tank or hot water tank. That's a convenient place for extra wind energy to go, with no waste. If the water gets too hot, invite a neighbor over for a shower. You can easily size this heater probe with the procedure given here (see box).

A final, important feature is a backup electrical generator for extended windless periods. Folks who travel about in large motorhomes know all about these generators. Some are noisy; others are not. They come with pull-starters or electric starters. The example shown here has a second load monitor wired to the starter. The monitor senses a very low voltage condition indicating a dying battery. It then starts the generator, or rings a bell to signal you to do the same. If you have to start the generator yourself very often, you are likely to develop interesting conservation practices.

Besides the losses that occur in the generation and storage of electrical energy, important losses occur in transmission. Make sure that wiring runs are as short as possible, wiring patterns are neat, electrical connections are sound, and the wire itself is sound and well-insulated. Also of critical importance is the diameter or gauge of the wire. Wires that are too small for the currents being carried will warm up excessively and waste energy. Use the procedure given in Appendix 3 to calculate the wire sizes necessary for your system. Then use one size larger than calculated. Wires tend to be permanent fixtures, while loads almost always grow.

What else can be done to enhance efficiency in a wind-electric system? Quite a bit, actually—especially in the area of load management. I've already talked about your electrical energy budget, and how to estimate it. In the course of studying your load characteristics, you will probably notice that some of the loads tend to crowd together, or be on at the same time. Through the magic of "peak shaving" you can cause loads to unbundle, lowering the peak power drain on your wind system (or on Edison, for that matter). This doesn't mean you will actually save energy in a measurable quantity. If you unplug the refrigerator while the toaster is running, the refrigerator will make up for the loss as soon as it's plugged back in. However, your batteries will last longer, and

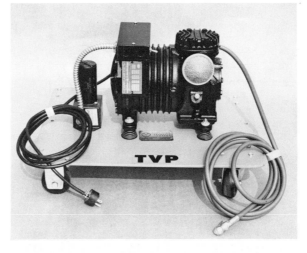

Above: One possible backup generator for a wind-electric system.

Right: A complete DC wind-electric system, including DC and AC loads, and back-up power source.

in the process of zipping around unplugging things, you might actually find areas for improvement. Some usage might be discarded simply to reduce the number of things that must be unplugged each morning; others will not be so obvious. But, during the effort, you'll probably be a lot more conscious of your load and end up using less energy.

Wind-Electric Water Pumps

Pumping water with electric well pumps is a common practice throughout the world. But, using a windcharger to run water pumps

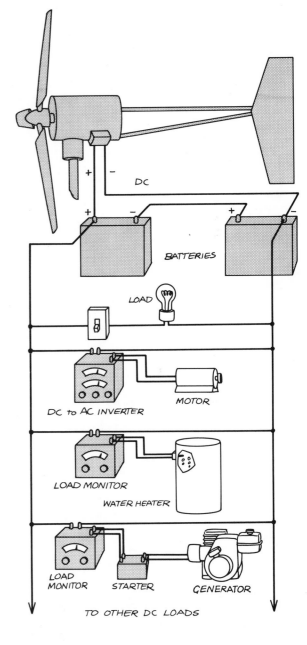

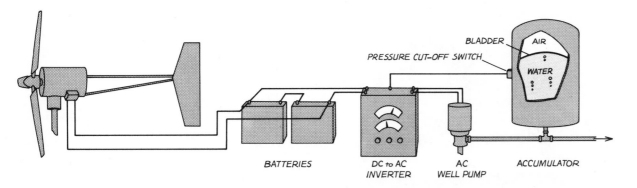

BLADDER AIR

PRESSURE CUT-OFF SWITCH

WATER

BATTERIES   DC to AC   AC   ACCUMULATOR
               INVERTER   WELL PUMP

Wind generator coupled to an electric well pump. The battery storage and inverter supply AC power as needed by the pump.

is not so common. Electric pumps for residential or small farm applications are usually sized from a half to a few horsepower and run on 120- or 220-volt alternating current. However, most small windchargers generate direct current that cannot be used directly with an AC pump. Also, well pumps fill an accumulator tank that holds a fixed volume of water at normal pressure. They then cycle on and off on demand to maintain accumulator pressure. If the accumulator goes bad, the pump cycles on every time a faucet leaks a few drops of water. If the accumulator is functioning properly you might get half-way through a shower before the well pump comes on.

The diagram above shows how you might wire a wind generator to a demand well pump. An inverter converts the DC from the wind machine to the AC voltage required by the pump.

This water pumping system is not really very compatible with wind power. The wind cannot be trusted to blow on demand, so batteries are required to bridge the time span between wind availability and water demand. Occasionally, but not often, the wind-generated electricity can be used directly by the pump—whenever the wind coincides with demand. The best batteries available to small-system users are only 60 to 80 percent efficient, and up to 40 percent of the wind energy is lost as waste heat. By contrast, whatever water you pump into a tank you usually get back one way or another. Water tank storage is virtually 99 percent efficient (some water usually leaks or evaporates).

The option to save up to 40 percent of the energy produced by a windcharger makes pumped-water storage systems much more desirable than battery-storage demand pumps. Here, you couple an AC well pump directly to a windcharger through an inverter— without using battery storage. The pump may be plumbed directly to a pond or irrigation ditch, or it may fill a storage tank. The storage tank will then serve the demand load. If the tank is high enough above the water user, no further pumping will be required.

A problem inherent with this system is load matching. The motor of an AC pump has a narrow range of rpm, and the well pump will occasionally supply the motor with a load that exceeds the power available from the windcharger. This results in a stalled motor; no water is pumped even though the windcharger is supplying power. Here, batteries would have made the difference.

Instead of an AC motor to drive the well pump, you might well consider a DC motor with a voltage and power rating compatible with the generator in the wind machine. An inverter is not required, but you might add a small electrical circuit (control box) to switch the pump off when too little power is available to pump any water. The AC motor considered earlier is a constant-power device that spins at a fixed rotational speed and drives the pump at a fixed rpm so that a virtually constant pumping rate is achieved. The DC motor, which is likely to be a converted traction motor from a golf cart, will spin at an rpm governed only by the power available to it. But a DC motor is not compatible with jet pumps (which many modern small wells use) because they are designed to spin at a fixed rpm. It is compatible with a positive displacement pump, such as a piston pump.

You might require a hybrid system that combines elements of most of the electrical applications diagrammed so far. A typical wind-electric system is not dedicated entirely to water pumping, although such a system is possible. The lower drawing shows a hybrid or flexible wind-electric system with major emphasis on water pumping. Note that the added control box uses the water level in the storage tank as one of its inputs. Another input might be power available from the windmill. The control box could sense a need for water in the tank, but hold off drawing from the batteries until the windmill

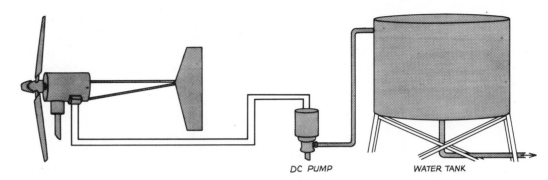

A simple wind-powered electric pumping system. The only storage is a large water tank.

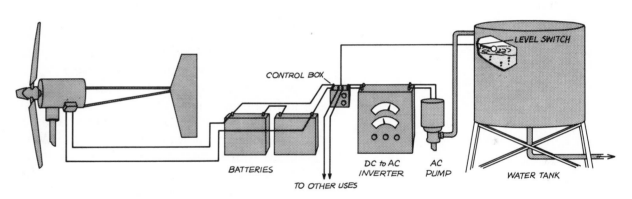

A complete wind-electric water-pumping system. The well pump is just one of several loads powered by this windcharger.

is delivering enough power to pump water. The control box could be set to sense a crisis water level in the tank and draw current from the batteries, or ring a bell to warn the owner. The controller could be programmed so that if more wind power was available than the batteries needed, the pump would bring the water levels up to a "full reserve" level—taking advantage of the occasional excess power a wind system might produce. Such programmable intelligence, when coupled with several wind energy applications, leads to higher overall system efficiency.

Chapter 2 described a novel technique for generating electricity and pumping water—the Bushland, Texas, experimental system being tested by the U.S. Department of Agriculture. This particular scheme uses the full capabilities of the electric well motor. If no wind power is available during pumping, the motor does the entire job, spinning at about 1,750 rpm. When the wind increases, an electronic control circuit releases the windmill brake, and a small starter motor spins the wind turbine up to speed. At that point, a clutch engages to couple the torque from the turbine to the torque supplied by the electric motor. Motor rpm increases as more wind power becomes available. When motor rpm reaches about 1,800, no electricity is needed from the utility. As the rpm increases above 1,800, the motor turns into a generator, pumping juice back into the utility lines.

A typical synchronous application for this type of wind-electric water pumping would be an industrial or agricultural water pump that drives a large irrigation system or an industrial process plant. Such a pump is probably a 220-volt or 440-volt motor from ten to several hundred horsepower in size. An Idaho potato-farmer uses a 400-hp electric pump to lift water 600 feet out of the ground into a holding pond. Four 14-hp pumps

supply water to his center-pivot irrigation systems, which are used to irrigate potatoes. The utility lines already wired to the pumps now provide all the necessary power, but a constant-frequency, synchronous wind generator could be wired to operate together with, or in parallel with, the utility lines. It is unlikely that a 400-hp wind system is within this farmer's budget, but any wind power supplied will offset the power required from the utility lines. Any or all of these pumps could be coupled to a wind turbine in an arrangement similar to the Bushland, Texas, machine.

Wind Furnaces

A wind furnace is a wind-powered heating system. Low-grade heat is needed for domestic purposes and for agricultural and industrial processes. Hot water can warm a room in winter, and provide a hot bath or sterilize a milking parlor, and wash manufactured parts before painting. Hot air warms rooms and barns and can dry parts after washing. Solar heat is rapidly replacing gas, oil, and coal-fired heat sources in many domestic applications, and in some industrial and agricultural processes. One could reasonably expect to apply wind energy to heating—perhaps as a complement to solar energy.

Wind furnaces can be used to supply the extra heat required for home climate control in regions where it is both windy and cold. Many parts of Eastern Canada, the New England coastal regions, and the Great Plains states and provinces are likely areas for important wind-furnace applications.

Space heat requirements of a building depend on the temperature difference between the indoor and outdoor air. Heat flows from warm to cold, and when this difference is large, serious heat losses ensue. Insulation, caulking and weatherstripping are typical measures to minimize these losses. But the overall heat loss can be severely increased by cold winds. The rate at which heat leaks out is increased by wind chill and by higher than normal infiltration—the process where cold air sneaks in through cracks and small openings. Wind increases infiltration by building up air pressure against the house, pushing more cold air inside. Wind chill is an increased conduction of heat away from the outside surfaces of the building. The wind carries heat away from these surfaces much faster than normal. Because of greater complexity, infiltration loads are much harder to predict than wind chill heat loads.

The beauty of using a wind furnace for extra winter heat is that it works hardest when it is needed most. A solar heating system or wood stove can provide your base heat requirements with a small wind furnace to provide peak load heat. This approach limits the need for energy storage—hence it's a cheaper system.

Your wind furnace can be used to generate heat for just about any purpose; the

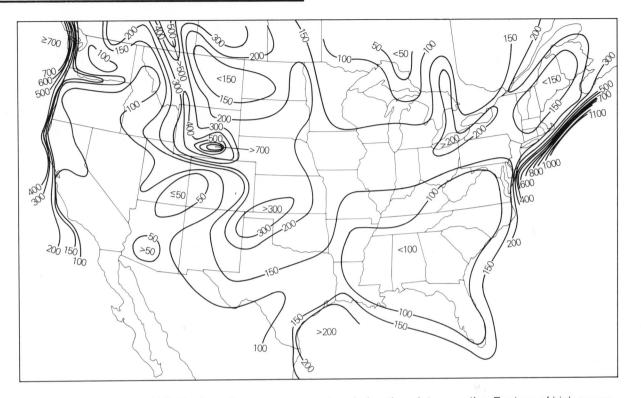

Average wind power available, in watts per square meter, during the winter months. Regions of high power are good areas for wind furnaces.

concept is not limited to space heating. However, space heating uses lower temperatures than most other processes (80–100°F, compared with 100–150°F for domestic hot water, and higher for many industrial processes). Lower temperatures are easier to reach and easier to keep. If you carefully size your storage tank and use plenty of insulation, a wind furnace can provide a fair portion of all your heating needs.

heat. Or a pump could be used to splash water against the tank walls. Similarly, the windmill can drive an air compressor. Compressed air will rise in temperature while it is being compressed, and as it is released through a turbulent outlet. Windmills for these purposes must develop high torque at low rpm; so high solidity is necessary.

A windmill can also drive a heat pump directly. The heat pump can be suitably designed to produce low temperatures as well as high. Thus, some of your refrigeration needs, as well as heating needs, can be satisfied with a heat pump. Such heat pump systems are already available for conventional and solar energy sources, and can be readily adapted to or designed for wind power.

Perhaps the most versatile heat generation method is to use a windcharger to drive an electric-resistance heater. Such a system (the University of Massachusetts Wind Furnace project) was described in Chapter 2. Electric baseboard heaters or heater probes for water tanks are very common. Wiring these devices to a windcharger is a simple matter, requiring only a load monitor to prevent overloading the windcharger in low winds. This versatile system can also provide electricity for other purposes when it's not needed for heat. Such a wind furnace is ideally suited to installation in tandem with a solar heating system, where a small amount of electricity is needed for solar pumps, controls, or fans, and the rest can be applied to heat.

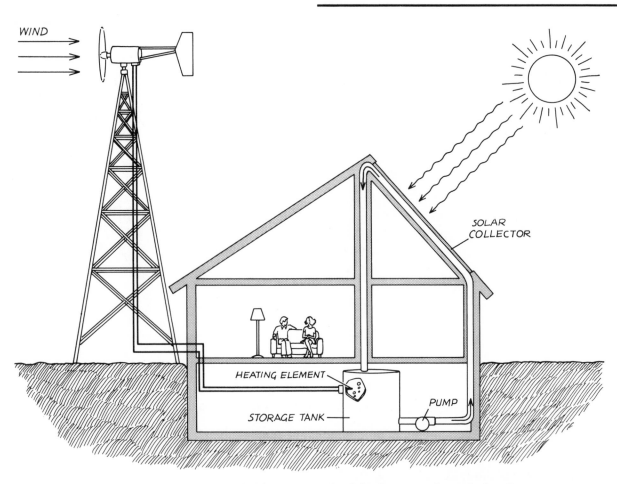

A wind furnace can be linked to an active solar heating system.

Wind energy can be converted to heat in several ways. If a windmill drives a paddle that splashes water around inside a closed, insulated tank, the water will be gradually warmed. Almost all of the mechanical energy delivered by the paddles is converted to

Building Heat Loss

Space heating is needed because heat flows from the warm, comfortable interiors of a building to the cold outdoor air. The rate at which heat must be added to a building should equal the rate at which heat leaves the building, or discomfort ensues. Calculations for this heat loss are not particularly complex, but they are tedious. They can be done with a cheap pocket calculator and an afternoon of concentration. Often they must be done to satisfy building code requirements before a new dwelling can be built. Complete heat loss calculations are too detailed to present here, but several good methods have been used to prepare a graph to help you estimate the heat loss of dwellings. This graph describes the heat loss rate Q for typical single-family dwellings ranging from poorly built, uninsulated to well-built, highly insulated structures. However, this graph by no means illustrates the best or worst you can do.

The heat loss rate Q is presented in units of Btu/DD/ft^2, or British Thermal Units per degree-day per square foot of floor area in the house. To calculate the total heat loss per month, you need to know the number of degree-days at your locale. This is a climatic variable that indicates how often and how much the average outdoor temperature falls below 65°F. Tables of degree-days are presented in Appendix 2.5. Then the monthly heat loss of a dwelling can be estimated using the formula:

$$Heat\ Loss = Q \times DD \times FA,$$

where DD is the number of degree-days for the month in question and FA is the heated floor area in square feet. The value of Q should be estimated using the graph here;

the heat loss will be given in units of Btu. To get the heat loss in kilowatt-hours, divide this answer by 3414.

Example: A house with 1000 square feet of heated floor area is located in Las Vegas, Nevada. If the house has R-11 insulation in the walls and R-19 in the ceiling, how large is the heat loss during the month of January?

Solution: From Appendix 2.5, there are 688 degree-days in Las Vegas during the month of January. Assuming this to be a windy month there, you can read Q = 11 along the 15-mph wind line in the graph. Then,

$$Heat\ Loss = 11 \times 688 \times 1000$$

$$= 7,568,000\ Btu.$$

Dividing this number by 3414, the heat loss equals 2,216 kWh—the amount of electrical energy needed to heat this house during the entire month. It would take a 3-kW wind generator operating constantly throughout the month to supply this much energy.

Suppose that only the extra heat loss caused by the wind itself is to be supplied by the wind furnace. This heat loss can be estimated by using the difference in Q between the windy (Q = 11) and no-wind (Q = 7) conditions in the formula:

$$Heat\ Loss = 4 \times 688 \times 1000$$

$$= 2,752,000\ Btu,$$

or 806 kWh. This amount of electrical energy could easily be supplied by a wind generator during a month with an average windspeed of 15 mph.

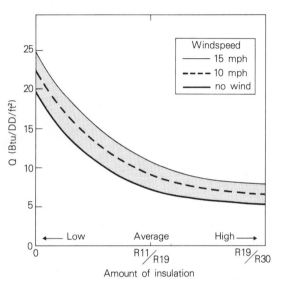

Wind machine installation using a gin pole. This Kedco machine was tilted up fully assembled with the tower.

Installation

Planning a wind system involves a complete understanding of the requirements of each component. With wind-powered water-pumpers, the plumbing may require freeze protection—often accomplished by burying pipes below the frost depth. Lightning protection must be considered in all installations. Other questions concern space. Do you have enough room to complete an installation and perform the necessary maintenance and repair operations? Do you have a cool, well-ventilated, well-protected area for batteries, electronic equipment, wiring and such? One of the greatest nuisances might send you off to buy 6-foot-high fencing, gates, signs, and a shotgun. Time and again, I've found complete strangers climbing my tower as though it had been installed just for that purpose!

The major points to consider in any installation are the following:
- Safety
- Efficiency
- Cost
- Environmental impact.

Your own situation will determine the order of importance for each of these points. The first three are discussed here; environmental impact is discussed in Chapter 7.

The dual consideration of cost and efficiency usually work against you in wind machine design—higher efficiency almost always costs more. On the other hand, shorter wire runs cost less and are more efficient.

But larger diameter wires are more efficient—and cost more. Taller towers get the rotor into higher winds, but cost much more and make installation and maintenance more difficult.

Planning a safe installation means understanding all the loads involved in supporting your wind machine aloft and selecting appropriate tower and foundation designs. Safety goes a bit further than that, however. It is not an easy task to erect a giant tower, especially in a tight space. It's even more difficult to work aloft with a heavy, cumbersome wind machine. Such a feat is a lot like overhauling a diesel-tractor engine 60 feet in the air.

Plan your installation allowing for enough room to tilt the tower up. This process may require a winch, a truck, ten friends and a gin pole. A gin pole is simply a support that allows the rope or cable you use for lifting the tower to start the lifting process from above the tower. Gin poles need not be very tall—perhaps one-third to one-half the height of the tower. As the tower rises under tension from the rope, passing over the gin pole top, the ability of that rope to continue raising the tower increases, and the need for the gin pole decreases. Once the tower is about halfway up, the gin pole won't be needed any more. The rope can finish the job directly. But be careful. Extra bracing may be needed to prevent tower collapse during this delicate operation.

You have several options for getting the wind machine aloft:

- Tilting it up with the tower
- Hoisting it aloft fully assembled
- Hoisting it aloft partially assembled
- Hoisting individual components for assembly aloft.

The first option is generally the easiest, unless your machine is very heavy. The last is time-consuming but reduces the need for hoisting equipment. You should consider this point carefully when designing your wind system. An appropriate design configuration would include a lightweight carriage structure with a built-in hoist that would rise with the tower. The built-in hoist would then bring up each component. For a small machine, the carriage can be built as a shell, thus doubling as a protective cowling.

Lightning Protection

A final, very important safety feature is lightning protection for your wind machine. Any wind machine is a prime target for lightning because it's usually the tallest metal object around. The map of thunderstorm frequency presented here shows the average number of days per year with thunderstorms over the U.S. and Southern Canada. From this map you can see that the Rocky Mountains and the southeastern United States are the two principal areas of major thunderstorm activity. Anywhere from one strike every other year to four strikes per year might be expected at or near your site.

Just how big is a typical lightning strike? Lightning current typically peaks at 20,000

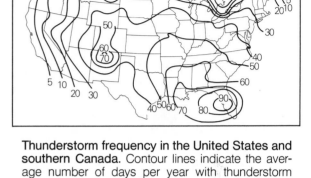

Thunderstorm frequency in the United States and southern Canada. Contour lines indicate the average number of days per year with thunderstorm occurrence.

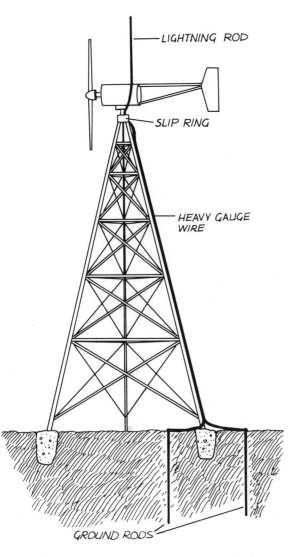

Properly grounded, a lightning rod establishes a cone of protection over a wind machine. The slip ring allows the machine to rotate without getting tangled.

amperes in a 1-microsecond pulse. About 2 percent of the time it peaks at 100,000 amperes in the same time interval. Clearly, you cannot afford to overlook lightning protection. Without an adequate electrical path to the ground, lightning will damage your wind machine with electrical heating, magnetic forces, or general mayhem. Corona balls—glowing balls of ionized air—have been seen to enter an electrical conduit by flowing along a wire. When a lightning ball enters an electrical box and discovers it has no way out, it responds like any irate prisoner—by destroying the contents of its confinement cell. On a less dramatic level, the very small static discharges associated with "electrical air" during a thunderstorm can easily zap transistor circuits and fuses. Over a long period of time, they can burn spots in windmill bearings or wiring.

A lightning rod is a device that provides a path for a lightning strike to reach ground. Most wind machines act as their own lightning rod. Some farmers keep their old windchargers aloft for the sole purpose of providing protection from lightning strikes on a nearby home. But this means that the massive current can flow directly through the delicate machinery. An alternative is to install a special lightning rod on the wind machine itself, providing a direct path to ground for the lightning current. Such an approach has been used on the wind machine that powers the wind furnace at the University of Massachusetts.

Whether or not a lightning rod is used,

suitable protection should be included in the wind machine design and installation. All bearings and shafts should have brushes that pass electric current *around* bearings, rather than through the bearing. This will prolong bearing life. All transistor circuits should be installed at the base of the tower or elsewhere, not aloft. Such a practice makes it much easier to replace zapped parts. Finally, wiring and grounding ought to be installed according to the building codes for lightning protection. The real key to success in any installation is a good electrical connection to the earth. In dry soils, this means many ground rods sunk 8 to 10 feet, or even more. In moist soils, you need not take as much care.

Comments From The Real World

Wind machines can be very useful in generating some of the power you need. They can also be quite dangerous, if improperly installed or maintained. After ten years of designing, testing, installing, and owning many different wind machines, I have found that most of these dangers appear only when people are around. Ropes break or knots loosen while tilting up a tower. Nuts, bolts, and tools fall from the tower top while someone is up there working. Worse yet, people often climb towers to work on their machines when they are tired and not really up to the concentration required.

The following tips are derived from my own experience with dozens of installations:

1. An installation takes more than twice as long to perform as the least optimistic guess would have it. Start early, or arrange for the project to be split into several tasks that can be performed on separate days.

2. Wind at the top of a tower is *always* stronger than at the bottom. Always make sure that blades cannot begin spinning before an installation is complete.

3. Really plan the up-tower operations carefully. Know the pocket in which each tool is kept, or who will perform which task, and when. Trying to choreograph the dance routine of a 400-pound elephant 60 feet aloft can be a difficult, if not downright dangerous, task.

4. Figure that tying a rope knot will lower that rope's strength to half its advertised value.

5. Try to avoid "fire-em-up-itis." Don't be in a hurry to let the machine spin under wind power. Let its first run occur during a very mild breeze. Save the full-power runs until you are certain that the rotor spins smoothly and *all the bolts are tight!*

6. Always wear a hard hat when working below somebody else.

7. Always wear a climbing belt and attach it firmly to a strong part of the tower.

8. Try not to look down!

7
Perspectives

Economic and Social Issues
of Wind Power Use

There are many factors other than energy needs and available technology that influence the decision to purchase or build a windpower machine. Legal and social constraints on the selection of wind energy systems stem from age-old questions concerning the rights and obligations of citizens. Financial issues seem to be the greatest barrier to widespread use of wind power, but as the prices of fossil fuels increase, wind systems will become increasingly competitive. Federal and state projects and tax incentives are rapidly testing and penetrating these institutional barriers—helping to increase the use of wind energy.

Wind Power Economics

I often hear people claim that wind power is "free." Others ask, "Will the wind power my house?" By now, you should have a good idea of what it takes to power your house, and you probably understand that although the wind blows whether you use it or not, harnessing the wind is definitely *not* free.

Then what does wind energy cost? Lots of folks simply add up all the costs involved in a complete wind system installation and stop right there. Similarly, lots of people simply ask how much a new house costs and don't look any further. But there's a new trend in cost assessment. It started with automobile purchases and is spreading to appliances. Soon it will reach home buying, and at the consumer level it's beginning to penetrate energy system purchases. That trend is toward serious consideration of the *life-cycle costs* of a system, house, appliance or car. The life-cycle cost includes not only the initial price of an installation but also its maintenance, fuel costs, and interest on money borrowed or spent. In home purchases, passive solar homes usually have a higher initial price but produce, at little or no operating cost, much of their own comfort-control energy. The monthly cost of owning such a home is usually less than that of a conventional home purchased at a lower initial price.

Similarly, two wind generators, each rated at 4kW, might each sell for a different price. The more expensive unit has a larger rotor diameter than does the cheaper unit. Some people would be inclined to buy the cheaper unit just to minimize their initial investment in a wind system. But which machine is actually cheaper? To answer that question you must determine several factors.

First, you need the total sales price of each completely installed system. You must add up the cost of everything from building permits to concrete foundation to wiring and testing. This is the so-called "first cost" of the system.

Next you need an accurate estimate of the amount of wind energy each system will produce at your site in a year's time. If the only data you have for the site is the mean annual windspeed, use a Rayleigh distribution (Chapter 3) for that speed to get the number of hours the wind blows at each

Wind Energy Costs

A complete economic analysis of wind energy versus other energy options requires an estimate of all costs incurred over the life of the wind system. You also need an estimate of the energy it produces per year. The annual cost of that energy can then be compared with the cost of conventional supplies.

To perform such a life-cycle estimate of your wind system costs, you need to know the following:

- *Purchase price of equipment, in dollars*
- *Installation costs, in dollars*
- *Annual maintenance costs, in dollars*
- *Annual insurance costs, in dollars*
- *Other annual costs, in dollars*
- *Resale value of equipment, in dollars*
- *Annual interest rate paid, in percent*
- *Expected system lifetime, in years*
- *Annual energy yield, in kilowatt-hours*

If you don't know all these quantities exactly, try to estimate them as best as possible. System lifetime, for example, is anybody's guess.

Start your analysis by adding up the purchase price and installation costs. Then multiply the sum of all the annual costs by the expected lifetime of the system, and add this product to the total above. Finally, subtract the estimated resale value of the equipment. Exclusive of financing costs, this result is the total cost of your system over its expected lifetime. To get the average annual costs, you divide this result by the expected lifetime and multiply by a factor that includes the costs of borrowing money. An example will help to illustrate.

Problem: *Suppose a wind system can be purchased off-the-shelf for $4,000, including wind generator, tower, batteries, wiring and controls. Suppose that installation of the system costs another $1,000. Based upon discussions with other owners of that model and with the manufacturer, you can expect a system lifetime of 20 years, with annual costs for insurance and maintenance averaging $200 per year. You estimate a resale value for all the equipment to be only $500 at the end of this 20-year period. If the system is projected to yield 3000 kWh per year at your site, what are you paying per kWh for the energy it produces?*

Solution: *First add up all the costs and subtract the resale value:*

Purchase price	$4,000
Installation cost	1,000
Maintenance & insurance	4,000
	$9,000
Resale value	−500
	$8,500

To get the average annual costs of the system, you should include interest paid as an added expense. Using simple interest at 10 percent per annum, the average annual costs of this system are:

$$1.10 \times \frac{\$8,500}{20} = \$467.50 \text{ per year}.$$

When divided by the energy produced per year, this number gives you the unit cost of the wind energy delivered by your system:

$$\frac{\$467.50}{3,000 \text{ kWh}} = \$0.16 \text{ per kWh}.$$

This figure may seem high, but wait a minute. If you take the Federal tax credit on the capital costs ($5,000) of the installation, you save 40 percent, or $2,000. This credit reduces your average annual costs to $357.50 and the unit cost to $0.12 per kWh. (State tax credits may reduce this cost even further.)

windspeed. Then use the power curve for each machine to get the watts available at each windspeed, and multiply watts times hours to get the total energy (in watt-hours) available at that speed. Finally, add up all the watt-hours available; this total is a rough estimate of each wind generator's annual energy yield.

Now you have the first cost and an estimate of the total energy yield, which is the annual yield times the expected lifetime of the system. If you stop right there and divide cost by energy yield, you get a very rough indication of the cost in dollars per watt-hour, or per kilowatt-hour, of energy produced. For a typical small wind machine, this might be anywhere from 10 to 30 cents per kWh. Depending on where you live, your current electric bill might be based on a price of 2 to 15 cents per kWh. But life-cycle cost analysis goes further.

Other costs that need to be included in your analysis are the costs of maintenance, insurance, financing, and taxes. Maintenance costs are difficult to estimate, but they might be covered by a maintenance contract with the installer. Insurance costs are easily identified by calling a broker. The financing costs of your installation are more complex. If you take money out of a savings account or other securities to purchase your wind system, you suffer a loss of interest. If you borrow the money, you usually pay an even higher interest rate. In either case, the cost of the money used is important to a fair economic assessment.

A helpful procedure to perform an annual assessment of the total system costs is given here (see box). In a life-cycle assessment, you add up all costs over the entire life of the machine, subtract the resale value estimate, and compare this total dollar figure with the savings in energy consumption costs over the expected system lifetime.

Among the many changing aspects involved in estimating your life-cycle cost are the federal, state, and perhaps even local government incentives to help you decide to buy wind equipment. The primary incentives are tax credits; many states and the federal government have programs that help finance your wind energy system by reducing your income tax liability. Tax credits will not pay the whole price, but in some states, like California, they reduce the cost by about 50 percent. You take this reduction into account when calculating the first cost of your system.

Similar government programs are starting to make low or no-interest loans available for conservation and energy equipment. The electric utilities are beginning to get into the act here. Other incentives may show up in the form of property-tax relief, or a reduced assessment of property value for people owning wind equipment. Such incentives will lower the annual costs you associate with the wind equipment.

When you compare your wind system energy cost with the cost of conventional energy sources using the life-cycle costs, you must include the effects of inflation.

Conventional energy costs are rising rapidly as shown in the graph. Once you install a wind system, you begin paying for it in the form of interest payments, taxes, maintenance and other costs. Some of these costs will be affected by inflation, but the major costs will occur at a fixed rate. Eventually, conventional energy costs should exceed the cost of your wind system. After that happens, the wind equipment is saving you money. A good windy site can begin to save money within a few years. A poor site might never save money. Tax credits may hasten the break-even day, but the selection of a good windy site is the most important factor.

Legal Issues

The legal issues involved in owning a wind power system cover two important areas—your rights and your obligations. You have certain rights, granted by law, that can be obtained by agreement or contract or that come automatically with land ownership. Other rights may have to be obtained through the courts. Your obligations are to protect the health, safety, and welfare of others.

Suppose a family has purchased property in a windy canyon and intends to build a home and install a wind generator. The legal questions that come with purchase of the property should be determined early on. A title search will find any restrictions placed on that property. If a previous owner speci-

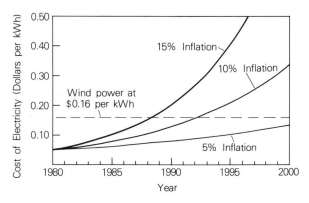

A comparison of wind electric and utility power costs. The wind power cost assumes a $5,000 purchase and installation fee, plus insurance and maintenance costs of $200 per year over a twenty-year lifetime.

The shaded states offer income tax credits or deductions for wind power equipment (1980).

style, and if the maximum structure height is limited to 20 feet, you might as well forget it.

Zoning ordinances are the next area of potential legal problems. The county may have certain restrictions that limit property to certain well-defined purposes—residential, agricultural, commercial, industrial, and so forth. These zoning ordinances may also specify architectural styles, building height limits, and other restrictions that affect a wind system plan. Zoning regulations are enacted for the purpose of protecting the "public health, safety, and welfare." They are usually administered by a zoning commission, planning department, or building inspector. You may apply for a variance whenever your project is at odds with an ordinance, and a hearing will be held to determine if the variance is to be granted. Potential wind-system owners typically have to apply for a variance if tower height exceeds the maximum height restriction.

Building codes are yet another source of problems. Before you can begin to lay the tower foundation, you will probably have to obtain a building permit. Typical building codes are the Uniform Building Code and the National Building Code; complete books on whichever code your county has adopted should be available at the local library. Building codes specify foundation, structural, electrical, and plumbing requirements. In some cases, your friendly building inspector may decide that the codes do not apply to your wind system. In most cases, however, they do. Your design will have to show, by

fically excluded the installation of any wind machine on the property, that would be the end of the wind system unless the new owner chose to test this restriction in court. Such a restriction is unlikely, but troublesome architectural restrictions often do exist. These can limit the height of a structure, determine its architectural style, or force the owner to submit to the whims of an architectural review board. It's hard to imagine a wind machine designed in Southern Colonial engineering calculations, that you have complied with the codes. In some cases, a registered professional engineer will have to check your drawings and calculations.

Often a deeply hidden restriction to your rights crops up because a wind system is an energy producer. In every state a Public Utilities Commission has licensed one or more utility companies to provide the energy needed by homes, farms, and businesses. In most cases, this public license amounts to a virtual monopoly. In some counties, the municipal water company has the right to install a water meter on your wind-powered water system—and charge you for the water.

The U.S. Federal Energy Regulatory Commission's recent adoption of the Public Utility Regulatory Policies Act of 1978 (PURPA) requires that utilities buy electricity from small power producers, including wind systems. This regulation now makes it possible for you to cogenerate electricity at your house or business, with full cooperation of the local electric utility. The Act specifies that utilities must buy this cogenerated power at their marginal cost of new power generation—which is a high rate these days. But the exact rate they pay will probably be the subject of debate and litigation for some years to come. However, your right to sell them excess power is clearly set forth in PURPA; it virtually eliminates any restrictions by a public utility on your ability to produce power.

Perhaps the most fundamental legal question concerns your right to the wind.

Surely whatever wind crosses your property is yours to use as you see fit. But what happens when your neighbor decides to build a high-rise structure just upwind of your machine? Zoning ordinances aside, you may not be able to stop such an action. Recently, a lot of legal debate has focused on sun rights and solar access—the rights to the sunshine that normally falls on a property or crosses an adjoining plot. A similar legal debate will eventually examine wind rights. For now, the problem is not highly critical; it is, however, something to think about in your site planning. If wind-rights appear to be a problem, you might try to obtain an easement. You usually obtain an easement for the rights to something concerning your neighbor's property. Easements are granted or sold for driveways, power line crossings, sewage lines, and other similar uses. A wind easement would restrict structure and tree heights on the part of your neighbor's property that is in your wind fetch area.

Along with your rights to harness wind power, you have certain obligations, most of which are concerned with the protection of life and limb. Many of these safety obligations will be satisfied in the course of complying with building codes. For example, preventing electrical shock is covered by electrical codes, while plumbing codes specify sanitation requirements for a water-pumping windmill that supplies domestic drinking water.

In my opinion the main area of concern

Flying rotor blades are one of the potential dangers of owning a wind system. Take care to plan for such hazards before installing the machine.

that may not be covered in your building codes is the attractive-nuisance value of a wind system. Install a swimming pool in your yard, and you will be required to fence it in. A swimming pool attracts children and creates a safety hazard. A wind system attracts everybody—not just children. And the safety hazard is made worse by the fact that people usually have less experience with windmills than they do with swimming pools. Your obligation is to protect trespassers from the safety hazard you create by installing a wind system. Check with an attorney, but expect to install at least a safety fence around the tower.

Your obligations extend to the protection of your neighbors from falling towers, flying blades, television interference, and other environmental damage. It may be that you install your tower so far from the property line that it cannot collapse into another yard. But you cannot so easily predict how far a broken blade will fly. Hence, you normally purchase some form of liability insurance to help cover any damages ... and keep your wind equipment in top shape so you don't need to use that insurance.

Social Issues

To a large extent, the social issues of technology are reflected in the laws, codes, and ordinances just discussed. The experience or mood of society, or needs of a group of people, help to guide the creation of laws that govern the applications allowed. If a group of wind energy systems is responsible for several accidents or injuries, a law will very likely be passed that governs the use of wind machines.

Social issues abound on a more personal level, too. Whenever you install a wind system, some other person or group of persons living or working nearby will react. During site evaluation you should simultaneously assess the social issues likely to occur at the site. Make sure your neighbor's reactions to your installation will be favorable. Neighborhood concern for safety will be first in importance. Next will be concern about any adverse effects your system will have on local television reception—unless cable television is used by all the neighbors. One company brought cable TV in with their wind generator to satisfy the neighbors. Noise and visual impact will also arise in discussions with the neighbors. Wind machines do not have to be much noisier than the wind that drives them; only poorly planned machines make substantial noise.

Looking over the entire range of tasks involved in planning your wind system, you may conclude it's a bigger task than you thought. Up to now, your only other option was to leave the entire job to a dealer and just purchase whatever equipment he recommends. Recent approval of the Residential Conservation Service (RCS) regulations now means that your utility must offer you an audit of your potential for conservation measures, solar energy, and wind energy. An RCS representative will help you with some of the initial planning steps; check with your local utility for specific details.

Because of RCS and PURPA—which allows you to cogenerate power—it's a sure bet that wind systems will soon become very attractive to a wide group of users. This growth in demand for wind equipment will stimulate new designs, more competition, and a wider selection of equipment. The various incentives to stimulate increased use of wind energy will take a firm grip on the market.

If you think your site is windy and you have figured out what it takes to power your house, it would make tremendous sense to contact your local utility RCS office and your state energy office. Ask them for information they may have accumulated on wind energy for your area. Also ask them for the forms necessary to qualify for any tax credits that may be available. Sort out your options and pursue the project carefully. The results will be more than satisfying.

How to Read a Graph

Graphs have been used throughout this book to simplify calculations. They also help to compress a lot of numerical information into a convenient visual form. Computations which would normally be difficult or laborious can be done easily with the help of a graph. But some readers may have trouble interpreting or using graphs; this Appendix is designed to help them. Here are a few illustrative examples.

Example 1: The curve in the first graph defines the relationship between two quantities, Value A and Value B. If Value A equals 8, what is the corresponding Value B?

Solution: Start on the horizontal scale ("x-axis") at 8 and move vertically (or draw a line) up to the curve. From this point of intersection, move horizontally (or draw a line) left to the vertical scale ("y-axis"). Thus, Value B = 4 in this example.

Example 2: Suppose you know that Value B = 2.5 in the first graph. What is the corresponding Value A?

Solution: From the vertical scale at 2.5 (halfway from 2 to 3), move horizontally right to intersect the curve, then drop down vertically to the horizontal scale as shown. Thus, Value A = 4 in this example.

Sometimes there is more than one curve on a graph that applies to your problem. Often, there are a series of curves, each corresponding to a specific value of one parameter. Here, you have to select the appropriate curve, or even add another curve to the graph. An example will illustrate.

Example 3: Look at the second graph. It contains two solid curves corresponding to two separate values of Value C, C = 10 and C = 20, in your problem. But you need to know what happens when Value C = 15. What do you do?

Solution: Value C = 15 is halfway between Value C = 10 and Value C = 20. Simply add another curve (dashed curve) roughly halfway between the two solid curves, and proceed as in the earlier examples.

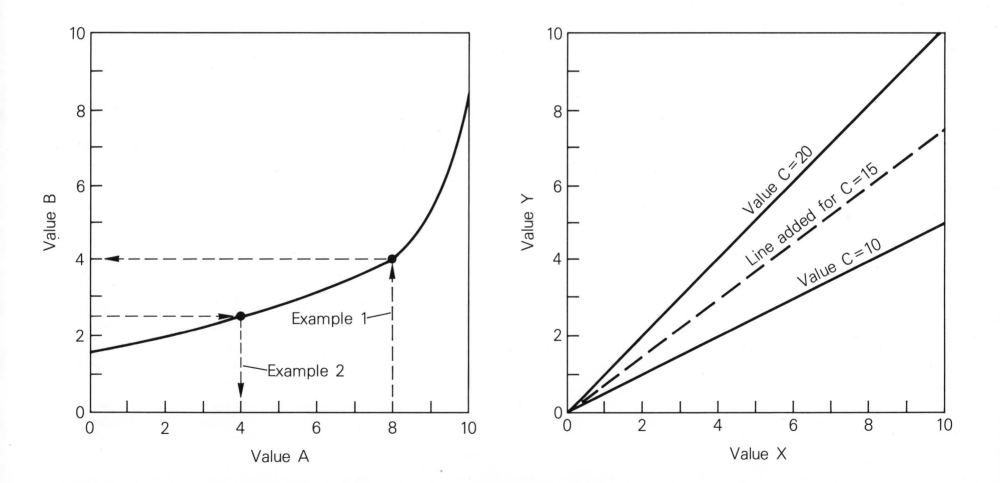

CONVERSION FACTORS

Multiply:	By:	To Obtain:
Atmospheres (atm)	76	Centimeters of Hg (0°C)
,,	1,033.3	Centimeters of $H_2O(4°C)$
,,	33.90	Feet of $H_2(39.2°F)$
,,	29.92	Inches of Hg(32°F)
,,	14.7	Pounds/inch^2
British Thermal Units (Btu)	252.0	calories
,,	777.65	Foot-pounds
,,	3.9275×10^{-4}	Horsepower-hours
,,	1,054.35	Joules
,,	2.929×10^{-4}	Kilowatt-hours
Btu/hr	4.20	calories/min
,,	777.65	Foot-pounds/hr
,,	3.927×10^{-4}	Horsepower
,,	2.929×10^{-4}	Kilowatts
,,	0.2929	Watts
Btu/ft^2	0.27125	Langleys (cal/cm^2)
Btu/ft^2/hr	3.15×10^{-7}	Kilowatts/meter^2
,,	4.51×10^{-3}	Langleys/min (cal/cm^2/min)
calories (cal)	0.003968	Btu
,,	3.086	Foot-pounds
,,	4.184	Joules
,,	1.162×10^{-6}	Kilowatt-hours
Calories, food (Cal)	1,000	calories
calories/cm^2 (Langleys)	3.69	Btu/ft^2
Centimeters (cm)	0.0328	Feet
,,	0.3937	Inches

CONVERSION FACTORS

Multiply:	By:	To Obtain:
Centimeters/sec (cm/sec)	0.0328	Feet/sec
,,	0.02237	Miles/hr
Cubic feet (ft^3)	0.0283	Cubic meters
,,	7.48	Gallons (U.S. liquid)
,,	28.32	Liters
Cubic meters (m^3)	35.315	Cubic feet
,,	264.2	Gallons (U.S. liquid)
,,	1,000.00	Liters
Feet (ft)	30.48	Centimeters
,,	12.0	Inches
,,	1.894×10^{-4}	Miles
Feet/minute (ft/min)	0.508	Centimeters/sec
,,	0.01829	Kilometers/hr
,,	0.01136	Miles/hr
Feet/second (ft/sec)	1.0974	Kilometers/hr
,,	0.6818	Miles/hr
Foot-pounds (ft-lbs)	0.001285	Btu
,,	0.324	calories
,,	3.766×10^{-7}	Kilowatt-hours
Gallons (U.S. liquid)	3,785.4	Cubic centimeters
,,	0.1337	Cubic feet
,,	231	Cubic inches
,,	0.003785	Cubic meters
,,	3.785	Liters
Gallons/minute (gpm)	2.228×10^{-3}	Cubic feet/sec
,,	0.06308	Liters/sec

CONVERSION FACTORS

Multiply:	By:	To Obtain:
Horsepower (hp)	2,546	Btu/hr
"	550	Foot-pounds/sec
"	745.7	Watts
Horsepower-hours	2,546	Btu
"	1.98×10^6	Foot-pounds
"	0.7457	Kilowatt-hours
Inches (in)	2.54	Centimeters
Joules	9.485×10^{-4}	Btu
"	0.7376	Foot-pounds
"	2.778×10^{-4}	Watt-hours
Kilograms (kg)	2.205	Pounds
Kilometers (km)	0.6214	Miles
Kilometer/hr (km/hr)	0.9113	Feet/sec
Kilowatts (kW)	3,414	Btu/hr
"	737.6	Foot-pounds/sec
"	1.341	Horsepower
Kilowatt-hours (kWh)	3,414	Btu
Langleys	1.0	calories/cm^2
Liters	1,000	Cubic centimeters
"	0.0353	Cubic feet
"	0.2642	Gallons (U.S. liquid)

CONVERSION FACTORS

Multiply:	By:	To Obtain:
Meters (m)	3.281	Feet
"	39.37	Inches
Meters/sec (m/sec)	2.24	Miles/hour
Miles (mi)	5280	Feet
"	1.61	Kilometers
Miles/hour (mph)	44.7	Centimeters/sec
"	88.0	Feet/min
"	1.61	Kilometers/hr
"	0.447	Meters/sec
Pounds (lbs)	0.4536	Kilograms
Square feet (ft^2)	0.0929	Square meters
Square inches (in^2)	6.452	Square centimeters
"	0.006944	Square feet
Square kilometers (km^2)	1.0764×10^7	Square feet
"	0.3861	Square miles
Square meters (m^2)	10.764	Square feet
Watts (W)	3.414	Btu/hr
"	0.001341	Horsepower
Watts/cm^2	3,172	Btu/ft^2/hr
Watt-hours	3.414	Btu
"	860.4	calories

Rayleigh Windspeed Distribution

The Rayleigh distribution is a one-parameter equation (see Chapter 3) that can be used to estimate the number of hours the wind blows at any particular windspeed. The single parameter in the equation is the annual average windspeed (the *mean* windspeed) at the site in question. The accompanying table presents values of the Rayleigh distribution for mean windspeeds ranging from 8 to 17 mph. Along each horizontal line in this table are the numbers of hours one would expect the wind to blow at each value of the windspeed listed in the leftmost column.

For example, suppose your site had a mean windspeed of 14 mph, and you wanted to estimate how often the wind blew at 23 mph. Just read down the column marked "14" at the top until you reach the horizontal line marked "23" at the left. You get the result "194," which means that the wind will blow at 23 mph approximately 194 hours per year at this site.

To get the result in terms of the *percent* of time each windspeed occurs, divide the table entry by 8760—the number of hours in a year. Thus, the wind blows at 23 mph for 194/8760, or 2.2 percent of the time.

RAYLEIGH DISTRIBUTION FOR VARIOUS MEAN WINDSPEEDS

Windspeed mph	Mean Windspeed, mph									
	8	9	10	11	12	13	14	15	16	17
8	784	731	666	601	539	484	435	391	353	320
9	716	697	656	605	553	503	457	415	377	344
10	630	644	627	594	554	512	470	431	395	363
11	536	578	585	570	543	510	476	441	408	377
12	441	504	533	536	523	500	473	443	415	386
13	351	429	474	494	494	483	464	441	416	391
14	272	356	413	446	459	458	448	432	412	391
15	204	288	353	396	420	429	427	418	404	387
16	149	227	295	345	378	396	403	400	392	380
17	105	175	242	296	336	361	375	379	377	369
18	73	132	194	250	294	325	345	355	358	355
19	49	97	153	207	253	289	314	330	337	339
20	32	70	119	170	216	254	283	303	315	321
21	20	50	90	136	181	220	252	275	291	302
22	12	34	68	108	150	189	222	248	268	281
23	7	23	50	84	123	160	194	222	244	260
24	4	15	36	65	99	134	168	197	220	239
25	3	10	25	49	79	111	143	173	198	218
26	1	6	18	37	62	91	122	150	176	197

RAYLEIGH DISTRIBUTION FOR VARIOUS MEAN WINDSPEEDS

Windspeed mph	Mean Windspeed, mph									
	8	9	10	11	12	13	14	15	16	17
27	.8	4	12	27	48	74	102	130	155	177
28	.4	2	8	20	37	60	85	111	136	158
29	.2	1	5	14	28	47	70	94	118	140
30	.1	.8	4	10	21	37	57	79	102	124
31	0	.5	2	7	16	29	46	66	87	108
32	0	.3	1	5	11	22	37	55	74	94
33	0	.1	.9	3	8	17	29	45	63	81
34	0	0	.5	2	6	13	23	37	53	70
35	0	0	.3	1	4	10	18	30	44	60
36	0	0	.2	.9	3	7	14	24	36	51
37	0	0	.1	.6	2	5	11	19	30	43
38	0	0	0	.4	1	4	8	15	24	36
39	0	0	0	.2	.9	3	6	12	20	30
40	0	0	0	.1	.6	2	5	9	16	25
41	0	0	0	0	.4	1	3	7	13	20
42	0	0	0	0	.3	.9	3	5	10	17
43	0	0	0	0	.2	.6	2	4	8	13
44	0	0	0	0	.1	.4	1	3	6	11

SURFACE FRICTION COEFFICIENT	
Description of Terrain	α
Smooth, hard ground; lake or ocean	0.10
Short grass on untilled ground	0.14
Level country with foot-high grass, occasional tree	0.16
Tall row crops, hedges, a few trees	0.20
Many trees and occasional buildings	0.22 - 0.24
Wooded country; small towns and suburbs	0.28 - 0.30
Urban areas, with tall buildings	0.40

Windspeed versus Height

If you know the surface friction coefficient α (Greek "alpha") at a wind site (see Chapter 3), you can readily estimate the windspeed at a given height h_B from measurements at another height h_A. The equation used to accomplish this feat is:

$$V_B = V_A \times \left(\frac{h_B}{h_A}\right)^{\alpha} ,$$

where V_A = the windspeed measured at height h_A,

V_B = the windspeed estimated at height h_B.

The surface friction coefficient α usually has a value between 0.10 (very smooth terrain) and 0.40 (very rough terrain). Typical values for α can be found in the table at left.

    The equation above can be used with various values of α to develop the series of "height correction factors" presented in the table on page 165. Here, α is listed at the top of each column, and the leftmost column lists the height in feet. To get the windspeed V_B at height h_B when you have measured the windspeed V_A at height h_A, use the following simple equation:

$$V_B = V_A \times \frac{H_B}{H_A} ,$$

where H_A and H_B are the height correction factors read from the table.

    These correction factors have been normalized to an assumed anemometer height of 30 feet above ground. They are also based on an assumption that the anemometer is not immersed within the layer of slow-moving air below the tops of trees or other nearby obstructions. For level terrain with few or no trees, height measurements start at ground level. When there is a grove of trees nearby, start all your height measurements at tree-top level.

Example: Suppose your anemometer is mounted 50 feet above ground level, but there is a grove of 30-foot trees just upstream. If your wind machine is to be mounted atop an 80-foot tower at this site, and the anemometer measures a mean windspeed of 10 mph, what is the mean windspeed at the machine height?

Solution: First you have to correct the machine height and the anemometer height for the grove of 30-foot trees. Subtracting 30 feet from each of the respective heights, the effective machine height (h_B) is 50 feet and the effective anemometer height (h_A) is 20 feet. Assuming a surface friction coefficient $\alpha = 0.28$ for wooded terrain, and reading down the column marked "0.28" in the table of height correction factors, we find that $H_A = 0.892$ and $H_B = 1.153$. Thus, the mean windspeed at the position of the wind machine is expected to be:

$$V_B = 10 \times \frac{1.153}{0.892}$$
$$= 12.9 \text{ mph}.$$

If there were no trees nearby (level country with only an occasional tree), and the anemometer measured 10 mph at 30 feet high, the mean windspeed at the 80-foot level would be, assuming $\alpha = 0.16$:

$$V_B = 10 \times \frac{1.170}{1.000}$$
$$= 11.7 \text{ mph}.$$

So the extra 50 feet of tower height gains you only 1.7 mph in mean windspeed over level terrain. But remember that wind power is proportional to the *cube* of the windspeed. The wind power available at the 80-foot level is 60 percent greater than that available at 30 feet.

	HEIGHT CORRECTION FACTOR, H								
Height	Surface Friction Coefficient, α								
(ft.)	0.100	0.140	0.160	0.200	0.220	0.240	0.280	0.300	0.400
10	0.895	0.857	0.839	0.802	0.785	0.768	0.735	0.719	0.644
15	0.933	0.908	0.895	0.870	0.858	0.846	0.823	0.812	0.757
20	0.960	0.945	0.937	0.922	0.914	0.907	0.892	0.885	0.850
25	0.981	0.975	0.971	0.964	0.960	0.957	0.950	0.946	0.929
30	1.000	1.000	1.000	1.000	1.000	1.000	1.000	1.000	1.000
35	1.016	1.022	1.025	1.031	1.034	1.037	1.044	1.047	1.063
40	1.029	1.041	1.047	1.059	1.065	1.071	1.083	1.090	1.121
45	1.041	1.058	1.067	1.084	1.098	1.102	1.120	1.129	1:176
50	1.052	1.074	1.085	1.107	1.118	1.130	1.153	1.165	1.226
55	1.062	1.089	1.102	1.128	1.142	1.156	1.184	1.199	1.274
60	1.072	1.102	1.117	1.148	1.164	1.180	1.214	1.231	1.319
65	1.080	1.114	1.132	1.167	1.185	1.203	1.241	1.261	1.362
70	1.088	1.126	1.145	1.184	1.204	1.225	1.267	1.289	1.403
75	1.096	1.137	1.158	1.201	1.223	1.245	1.292	1.316	1.442
80	1.103	1.147	1.170	1.216	1.240	1.265	1.316	1.342	1.480
85	1.110	1.157	1.181	1.231	1.257	1.283	1.338	1.366	1.516
90	1.116	1.166	1.192	1.245	1.273	1.301	1.360	1.390	1.551
95	1.122	1.175	1.203	1.259	1.288	1.318	1.380	1.413	1.585
100	1.128	1.184	1.212	1.272	1.303	1.335	1.400	1.435	1.618
105	1.133	1.192	1.222	1.284	1.317	1.350	1.420	1.456	1.650
110	1.139	1.199	1.231	1.296	1.330	1.365	1.438	1.476	1.681
115	1.144	1.207	1.240	1.308	1.343	1.380	1.456	1.496	1.711
120	1.149	1.214	1.248	1.319	1.356	1.394	1.474	1.515	1.741
125	1.154	1.221	1.257	1.330	1.368	1.408	1.491	1.534	1.769
130	1.158	1.228	1.264	1.340	1.380	1.421	1.507	1.552	1.797
135	1.162	1.234	1.272	1.350	1.392	1.434	1.523	1.570	1.825
140	1.167	1.241	1.280	1.360	1.403	1.447	1.539	1.587	1.851
145	1.171	1.247	1.287	1.370	1.414	1.459	1.554	1.604	1.878
150	1.175	1.253	1.294	1.379	1.424	1.471	1.569	1.620	1.903

Wind Power Tables

This appendix lists monthly average wind power data for 742 stations in the United States and southern Canada. These data have been extracted from the report *Wind Power Climatology in the United States*, by Jack Reed (see Bibliography); they represent averages over at least 20 years of wind measurements made at airports and weather stations. Average monthly and yearly wind power available at these sites is listed in units of watts per square meter; multiply by 0.0929 to convert these numbers to watts per square foot.

These data have *not* been corrected for the various heights above ground level of the anemometers used for each measurement. They also reflect many possible distortions in the wind patterns caused by natural terrain features and nearby buildings and trees. Thus, no particular set of these data can be blindly accepted as representative of a particular region. They do, however, provide a rough idea of the wind power available and the monthly variation in wind power at a large number of sites. By comparison of your measurements with those of nearby sites in these tables, you can obtain a better idea of the patterns to expect at your own site.

The measuring stations are listed by region within each state or province; states are listed first in alphabetical order, followed by Canadian provinces. Each line of the table contains the following information:

1. State or province (obvious abbreviations);
2. Exact location of the measuring station;
3. International station number;
4. Latitude in degrees and minutes (3439 means 34°39′ N);
5. Longitude in degrees and minutes (8646 means 86°46′ W);
6. Average annual windspeed, or mean windspeed, in knots (To convert to mph, multiply by 1.15);
7. Twelve values of the monthly average wind power available, in watts per square meter;
8. Average of these twelve monthly averages.

The most commonly used abbreviations for the location (#2 above) are:

APT — airport
AFB — Air Force Base
AFS — Air Field Station
IAP — International Airport
IS — island
NAF — Naval Air Field
PT — point
WBO — Weather Bureau Office.

Try to contact the actual station for more detailed information about local wind patterns. Jack Reed's report is also worth a closer look. Besides the data presented in this table, this report contains rough windspeed distributions—the percentage of time each month that the windspeed was recorded in each of eight different speed ranges—for each station listed here.

	MONTHLY AVERAGE WIND POWER IN THE UNITED STATES AND SOUTHERN CANADA																		

| State | Location | Lat | Long | Ave. Speed Knots | Wind Power, Watts per Square Meter | | | | | | | | | | | | |
					J	F	M	A	M	J	J	A	S	O	N	D	Ave.
AL	Huntsville	3439	8646	6.6	80	109	118	87	48	37	29	31	56	50	78	87	66
AL	Foley	3358	8605	8.0	133	182	153	174	142	106	73	66	109	96	116	115	122
AL	Gadsden	3358	8605	5.8	84	104	114	99	43	34	25	21	40	45	63	63	61
AL	Birmingham APT	3334	8645	7.3	127	157	156	137	80	64	49	44	68	68	108	106	97
AL	Tuscaloosa, Vn D Graf APT	3314	8737	5.1	79	79	93	69	33	21	15	21	28	36	55	69	49
AL	Selma, Craig AFB	3221	8659	5.7	74	86	91	68	42	34	29	26	37	31	48	54	51
AL	Montgomery	3218	8624	6.1	74	90	85	68	39	35	34	26	39	36	51	62	53
AL	Montgomery, Maxwell AFB	3223	8621	4.8	58	67	69	51	28	25	20	19	27	26	39	45	39
AL	Ft. Rucker, Cairns AAF	3116	8543	4.7	40	50	55	41	23	18	12	12	19	19	29	35	30
AL	Evergreen	3125	8702	5.3	66	69	78	57	29	18	17	17	21	24	38	51	40
AL	Mobile, Brookley AFB	3038	8804	7.3	105	104	128	119	94	58	43	41	69	51	72	89	80
AK	Annette IS	5502	13134	9.5	320	264	216	199	110	97	71	77	128	297	324	319	199
AK	Ketchikan	5521	13139	5.8	59	53	42	52	47	37	38	44	43	67	75	75	57
AK	Craig	5529	13309	7.9	185	159	167	132	82	95	71	55	113	186	174	165	128
AK	Petersburg	5649	13257	3.7	26	40	37	41	32	23	22	22	24	29	22	21	33
AK	Sitka	5703	13520	3.5	109	26	34	42	27	23	22	14	34	44	46	93	37
AK	Juneau PT	5822	13435	7.5	119	134	123	127	95	70	60	67	108	170	157	159	115
AK	Haines	5914	13526	8.0	218	202	203	148	74	61	94	54	72	160	238	159	146

Adapted by Dr. Richard Schwind from Jack Reed, *Wind Power Climatology in the United States*

				Ave. Speed Knots	Wind Power, Watts per Square Meter											
State	Location	Lat	Long		J	F	M	A	M	J	J	A	S	O	N	D	Ave.
AK	Yakutat APT	5931	13940	7.0	177	144	114	100	90	71	56	64	96	181	183	169	114
AK	Middleton IS AFS	5927	14619	11.9	625	597	468	355	238	141	96	134	243	519	582	608	376
AK	Cordova, Mile 13 APT	6430	14530	4.4	46	48	42	41	37	23	18	17	32	53	47	48	36
AK	Valdez	6107	14616	4.3	72	28	75	41	36	16	13	7	7	45	100	172	53
AK	Anchorage IAP	6110	15001	5.9	61	95	48	61	108	76	61	52	46	38	38	50	61
AK	Anchorage, Merrill Fld	6113	14950	4.9	57	66	29	27	41	40	23	22	30	30	59	23	37
AK	Anchorage, Elmendorff AFB	6115	14948	4.4	46	60	50	41	40	34	24	22	26	30	46	33	36
AK	Kenai APT	6034	15115	6.6	96	109	94	66	61	63	56	54	53	83	85	80	74
AK	Northway APT	6257	14156	3.9	16	21	30	44	40	42	33	32	27	22	18	16	28
AK	Gulkana	6209	14527	5.8	45	88	85	105	111	98	83	100	95	76	48	40	81
AK	Big Delta	6400	14544	8.2	447	322	239	147	148	85	68	102	163	209	300	333	215
AK	Fairbanks IAP	6449	14752	4.3	10	16	25	37	50	44	33	29	28	22	13	10	27
AK	Fairbanks, Ladd AFB	6451	14735	3.5	10	17	24	28	38	35	23	29	23	24	12	9	23
AK	Ft. Yukon APT	6634	14516	6.7	30	41	64	81	91	84	86	81	74	52	31	31	64
AK	Nenana APT	6433	14905	5.1	68	44	48	45	46	34	27	26	33	42	45	44	42
AK	Manley Hot Springs	6500	15039	4.8	76	54	84	109	93	89	53	42	63	104	62	52	62
AK	Tanana	6510	15206	6.6	72	85	89	83	56	53	47	29	50	64	56	80	73
AK	Ruby	6444	15526	6.5	58	133	119	84	40	51	46	42	64	76	119	54	79
AK	Galena APT	6444	15656	5.4	56	69	66	77	51	53	48	61	61	59	59	49	59
AK	Kaltag	6420	15845	4.7	46	103	26	81	31	33	30	28	37	56	40	51	56

						Wind Power, Watts per Square Meter												
					MONTHLY AVERAGE WIND POWER IN THE UNITED STATES AND SOUTHERN CANADA													
State	Location	Lat	Long	Ave. Speed Knots	J	F	M	A	M	J	J	A	S	O	N	D	Ave.
AK	Unalakleet APT	6353	16048	10.5	520	502	336	191	112	96	116	146	175	234	395	376	265
AK	Moses Point APT	6412	16203	10.6	329	363	275	279	149	129	181	233	217	222	246	263	241
AK	Golovin	6433	16302	9.6	188	229	246	250	142	117	178	264	271	258	369	256	236
AK	Nome APT	6430	16526	9.7	328	308	228	225	153	119	117	162	189	230	263	238	217
AK	Northeast Cape AFS	6319	16858	11.0	468	263	246	347	239	137	218	240	288	462	632	387	328
AK	Tin City AFS	6534	16755	15.0	763	919	811	658	427	271	260	334	352	522	722	728	549
AK	Kotzebue	6652	16238	11.2	455	418	310	294	161	187	212	234	228	270	397	366	291
AK	Cape Lisburne AFS	6853	16608	10.5	432	268	335	266	227	210	303	216	266	432	444	333	314
AK	Indian Mountain AFS	6600	15342	5.4	115	113	88	58	57	37	30	36	52	86	95	104	70
AK	Bettles APT	6655	15131	6.3	28	44	57	62	66	62	44	38	43	43	43	47	48
AK	Wiseman	6726	15013	3.1	28	26	16	15	24	15	23	14	12	12	26	16	23
AK	Umiat	6922	15208	6.0	113	121	43	77	80	93	62	53	57	51	121	78	76
AK	Point Barrow	7118	15647	10.5	215	194	162	167	169	143	145	208	211	258	286	183	192
AK	Barier IS	7008	14338	11.3	512	468	379	279	216	145	123	208	287	470	486	425	341
AK	Sparrevohn AFS	6106	15534	4.7	69	73	108	76	47	35	36	41	54	63	74	82	63
AK	McGrath	6258	15537	4.2	13	27	29	37	39	35	34	35	32	24	15	12	27
AK	Tataline AFS	6253	15557	4.4	25	37	36	37	38	27	27	29	33	35	24	21	31
AK	Flat	6229	15805	8.1	206	266	205	150	116	100	81	108	143	168	185	184	172
AK	Aniak	6135	15932	5.6	51	59	63	59	49	37	27	34	41	47	47	42	40
AK	Bethel APT	6047	16148	9.8	229	258	224	166	125	108	110	137	140	158	185	211	171

					Ave. Speed Knots	Wind Power, Watts per Square Meter												
State	Location	Lat	Long			J	F	M	A	M	J	J	A	S	O	N	D	Ave.
AK	Cape Romanzof AFS	6147	16602		11.7	692	699	493	476	246	124	110	154	234	305	520	654	380
AK	Cape Newenham AFS	5839	16204		9.8	400	371	330	288	168	119	101	142	165	212	300	315	241
AK	Kodiak FWC	5744	15231		8.8	328	271	258	198	124	87	52	77	120	210	294	329	189
AK	King Salmon APT	5841	15639		9.2	250	260	235	180	182	138	92	139	156	180	230	206	191
AK	Port Heiden APT	5657	15837		12.9	576	564	493	361	289	273	225	381	466	51	439	565	429
AK	Port Mollor	5600	16031		8.8	158	168	171	195	135	81	108	144	164	222	260	219	172
AK	Cold Bay APT	5512	16243		14.6	736	731	699	580	506	465	428	507	462	606	652	631	573
AK	Dutch Harbor NS	5353	16632		9.6	355	376	295	223	135	125	69	105	169	390	419	266	233
AK	Driftwood Bay	5358	16651		8.0	204	203	154	148	115	72	88	77	71	120	161	182	131
AK	Umnak IS, Cape AFB	5323	16754		13.5	651	688	577	514	454	251	163	249	466	603	606	723	497
AK	Nikolski	5255	16847		14.0	538	560	532	566	437	321	239	283	361	634	732	662	482
AK	Adak	5153	17638		12.2	426	467	528	453	366	223	218	258	331	502	481	525	404
AK	Amchitka IS	5123	17915		18.0	1764	1517	1418	1062	653	448	405	457	740	1053	1165	1569	1025
AK	Attu IS	5250	17311		11.2	553	582	508	403	235	162	135	129	360	366	414	554	368
AK	Shemya APT	5243	17406		15.7	887	932	878	641	483	266	235	285	432	301	977	870	633
AK	St. Paul IS	5707	17016		15.0	758	867	684	518	355	207	175	282	399	693	691	791	547
AZ	Grand Canyon	3557	11209		6.2	38	43	49	71	66	55	35	31	57	58	44	28	49
AZ	Winslow APT	3501	11044		7.3	104	104	232	169	161	141	93	77	63	73	63	78	111
AZ	Flagstaff, Pulliam APT	3508	11140		6.4	71	70	96	95	93	86	40	33	52	56	69	69	69
AZ	Maine	3509	11157		8.9	132	186	218	253	240	224	111	68	116	178	139	158	151

MONTHLY AVERAGE WIND POWER IN THE UNITED STATES AND SOUTHERN CANADA

	MONTHLY AVERAGE WIND POWER IN THE UNITED STATES AND SOUTHERN CANADA																	
				Ave. Speed Knots	Wind Power, Watts per Square Meter												
State	Location	Lat	Long		J	F	M	A	M	J	J	A	S	O	N	D	Ave.
AZ	Ashfork	3514	11233	7.5	111	116	154	201	142	126	82	68	86	100	103	82	114
AZ	Kingman	3516	11357	8.9	126	156	172	203	153	166	126	99	102	124	115	107	138
AZ	Prescott	3439	11226	7.5	54	95	117	144	138	124	75	56	67	57	59	44	85
AZ	Yuma APT	3240	11436	6.8	55	62	68	77	71	69	93	77	45	40	55	51	62
AZ	Phoenix	3326	11201	4.8	16	28	34	39	37	35	43	31	28	24	22	17	29
AZ	Phoenix, Luke AFB	3332	11223	4.6	21	31	41	52	49	43	49	39	25	21	20	18	34
AZ	Chandler, Williams AFB	3318	11140	4.1	17	21	28	35	34	33	41	33	28	22	18	16	26
AZ	Tucson APT	3207	11056	7.3	71	59	69	90	87	73	75	54	62	78	82	72	74
AZ	Tucson	3207	11056	7.1	71	59	69	90	87	73	75	54	62	78	82	72	74
AZ	Tucson, Davis-Monthan AFB	3210	11053	5.7	48	48	57	63	56	60	51	35	42	40	43	45	49
AZ	Ft. Hauchuca	3134	11020	5.7	49	58	84	96	80	66	39	27	30	30	34	41	53
AZ	Douglas	3128	10937	6.4	84	94	166	143	128	86	61	46	47	63	62	75	87
AR	Walnut Ridge APT	3608	9056	6.0	93	81	103	104	58	44	27	22	28	43	64	75	62
AR	Blytheville AFB	3558	8957	6.4	85	106	108	111	66	41	26	25	36	39	67	71	65
AR	Ft. Smith APT	3520	9422	7.4	76	86	116	104	81	62	51	45	50	53	66	75	73
AR	Little Rock	3444	9214	7.6	82	91	105	96	70	58	46	46	48	50	73	71	70
AR	Jacksonville, Ltl. Rk. AFB	3455	9209	5.8	61	67	85	70	47	34	28	24	28	29	45	49	48
AR	Pine Bluff, Grider Fld	3410	9156	6.5	102	89	102	87	51	39	31	29	35	45	71	81	64
AR	Texarkana, Webb Fld	3327	9400	7.7	92	108	128	115	77	69	48	49	62	61	74	87	80
CA	Needles APT	3446	11437	6.7	108	125	128	112	108	98	67	67	58	78	113	124	97

				Ave. Speed Knots	Wind Power, Watts per Square Meter												
State	Location	Lat	Long		J	F	M	A	M	J	J	A	S	O	N	D	Ave.
CA	El Centro NAAS	3249	11541	7.7	98	126	171	208	225	189	80	73	79	86	98	76	127
CA	Thermal	3338	11610	9.1	66	79	103	149	191	153	125	114	119	92	76	63	111
CA	Imperial Bch.,.Ream Fld	3234	11707	5.9	48	51	54	54	52	45	35	32	30	31	43	42	43
CA	San Diego, North IS	3243	11712	5.3	32	41	56	56	49	41	33	32	35	31	30	32	39
CA	San Diego	3244	11710	5.4	30	33	40	47	47	40	30	29	27	25	22	21	31
CA	Miramar NAS	3252	11707	4.4	23	24	28	30	26	19	16	17	18	19	20	24	22
CA	San Clemente IS NAS	3301	11835	6.3	53	72	89	97	67	48	33	32	33	32	54	69	55
CA	San Nicholas IS	3315	11948	9.9	152	199	295	306	348	244	161	166	164	140	180	159	209
CA	Camp Pendleton	3313	11724	5.2	30	35	45	61	53	43	43	44	36	24	28	29	38
CA	Oceanside	3318	11721	8.0	129	122	108	82	67	59	49	49	64	66	96	116	87
CA	Laguna Beach	3332	11747	5.0	35	38	44	37	30	30	27	26	25	25	22	32	34
CA	El Toro MCAS	3340	11744	4.8	45	38	33	30	26	22	19	19	19	23	36	43	28
CA	Santa Ana MCAF	3342	11750	4.6	43	43	47	46	37	31	30	26	25	26	36	43	37
CA	Los Alimitos NAS	3348	11807	4.8	36	39	47	44	41	32	28	25	22	23	36	37	34
CA	Long Beach APT	3349	11809	4.9	27	40	45	48	43	35	34	32	31	27	29	26	35
CA	Los Angeles IAP	3356	11824	5.9	40	57	69	70	63	49	43	44	39	36	38	35	48
CA	Ontario	3404	11737	7.7	36	117	109	118	148	124	135	135	92	71	46	127	103
CA	Riverside, March AFB	3353	11715	4.4	35	43	40	44	49	51	52	49	37	28	29	32	41
CA	San Bernardino, Norton AFB	3406	11715	3.5	43	43	33	27	25	22	22	20	19	17	28	29	28
CA	Victorville, George AFB	3435	11723	7.7	99	134	170	183	163	135	87	85	74	70	87	90	118

MONTHLY AVERAGE WIND POWER IN THE UNITED STATES AND SOUTHERN CANADA

	MONTHLY AVERAGE WIND POWER IN THE UNITED STATES AND SOUTHERN CANADA																
				Ave. Speed	Wind Power, Watts per Square Meter												
State	Location	Lat	Long	Knots	J	F	M	A	M	J	J	A	S	O	N	D	Ave.
CA	Daggett	3452	11647	9.6	94	173	315	290	355	236	177	159	145	121	107	74	137
CA	China Lake, Inyokern NAF	3541	11741	7.1	121	156	238	249	225	186	124	126	113	124	103	93	155
CA	Muroc, Edwards AFB	3455	11754	7.9	90	118	187	206	236	230	155	131	99	87	82	83	141
CA	Palmdale	3438	11806	10.2	163	205	226	267	315	328	254	200	165	158	130	109	225
CA	Palmdale APT	3438	11805	8.8	121	146	233	234	234	229	173	141	107	104	113	132	163
CA	Saugus	3423	11832	6.3	105	128	88	96	96	108	101	88	67	76	105	90	89
CA	Van Nuys	3413	11830	4.6	105	82	66	50	43	21	22	19	18	22	90	69	49
CA	Oxnard AFB	3413	11905	4.4	63	56	49	46	43	26	20	19	19	31	50	76	41
CA	Point Mugu NAS	3407	11907	5.6	100	79	71	78	51	33	28	26	28	35	76	82	55
CA	Santa Maria	3454	12027	6.5	75	80	114	94	93	93	63	57	56	66	79	91	82
CA	Vandenberg, Cooke AFB	3444	12034	6.1	62	67	99	97	115	67	34	33	41	51	58	58	65
CA	Pt. Arguello	3440	12035	7.2	72	105	138	135	133	79	58	54	51	74	76	66	85
CA	San Louis Obispo	3514	12039	6.9	60	69	134	127	146	173	105	120	131	129	89	73	115
CA	Estero	3526	12052	4.3	83	66	69	76	60	50	22	31	42	47	44	77	53
CA	Paso Robles, Sn Ls Obispo	3540	12038	5.5	34	39	57	76	105	127	106	83	59	42	32	30	64
CA	Jolon	3600	12114	2.8	9	6	10	6	11	11	8	8	6	4	6	6	7
CA	Monterey NAF	3635	12152	5.0	30	33	45	48	51	45	35	32	23	21	20	30	35
CA	Ft. Ord, Fritzsche AAF	3641	12146	5.7	30	31	46	61	67	63	66	59	41	34	25	24	47
CA	Taft, Gardner Fld.	3507	11918	4.4	20	18	20	29	45	46	31	22	18	16	18	17	26
CA	Bakersfield, Meadows Fld.	3525	11903	5.4	27	33	46	55	69	65	47	43	34	25	24	28	41

				Ave. Speed Knots	Wind Power, Watts per Square Meter												
State	Location	Lat	Long		J	F	M	A	M	J	J	A	S	O	N	D	Ave.
CA	Bakersfield, Minter Fld.	3530	11911	5.0	26	31	38	49	61	73	38	25	22	21	19	25	34
CA	Lemoore NAS	3620	11957	4.8	21	30	38	40	45	47	35	29	25	27	18	19	30
CA	Fresno, Hammer Fld.	3646	11943	5.5	24	28	42	48	60	62	42	33	25	23	17	20	35
CA	Bishop APT	3722	11822	7.5	74	106	161	145	129	100	80	81	85	101	88	80	103
CA	Merced, Castle AFB	3722	12034	6.0	56	66	72	74	69	78	59	52	44	44	34	42	59
CA	Livermore	3742	12147	7.9	109	108	115	124	158	180	173	143	107	85	64	71	122
CA	San Jose APT	3722	12155	6.4	51	47	61	61	86	84	52	43	46	34	45	47	54
CA	Sunnyvale, Moffett Fld.	3725	12204	5.4	47	50	54	59	65	73	62	54	41	35	32	56	51
CA	San Francisco IAP	3737	12223	9.5	96	129	183	228	268	280	236	211	171	141	80	91	176
CA	Farallon IS	3740	12300	9.6	61	406	287	193	188	208	100	91	83	106	204	275	212
CA	Alameda FWC	3748	12210	7.4	92	94	122	125	129	124	99	87	65	65	69	81	93
CA	Oakland	3744	12212	6.8	52	75	77	92	101	98	74	69	57	50	40	51	71
CA	San Rafael, Hamilton AFB	3804	12231	4.8	51	52	54	52	50	50	39	39	30	34	32	51	45
CA	Fairfield, Travis AFB	3816	12156	10.7	114	153	176	232	347	488	577	481	332	182	106	91	270
CA	Point Arena	3855	12342	13.0	401	398	361	488	500	614	388	513	321	368	320	467	421
CA	Sacramento	3831	12130	7.8	145	145	126	118	116	128	92	83	64	70	61	123	95
CA	Sacramento, Mather AFB	3834	12110	6.0	117	108	89	69	63	69	56	45	38	46	59	88	72
CA	Sacramento, McClellan AFB	3840	12124	6.5	107	102	98	79	83	90	62	56	49	67	75	84	79
CA	Auburn	3857	12104	-8.4	106	148	109	76	77	64	65	65	64	57	69	67	83
CA	Blue Canyon APT	3917	12042	8.4	237	212	168	106	92	75	57	64	65	110	130	188	122
CA	Donner Summit	3920	12022	12.1	1100	619	729	269	266	226	173	168	154	439	579	645	463

Table title: **MONTHLY AVERAGE WIND POWER IN THE UNITED STATES AND SOUTHERN CANADA**

	MONTHLY AVERAGE WIND POWER IN THE UNITED STATES AND SOUTHERN CANADA																
				Ave. Speed Knots	Wind Power, Watts per Square Meter												
State	Location	Lat	Long		J	F	M	A	M	J	J	A	S	O	N	D	Ave.
CA	Beale AFB	3908	12126	5.1	75	59	64	56	49	52	31	29	34	39	43	62	50
CA	Williams	3906	12209	8.2	163	172	179	112	126	120	78	64	78	105	112	116	111
CA	Ft. Bragg	3927	12349	5.9	75	88	82	96	46	44	25	26	25	33	50	52	51
CA	Eureka, Arkata APT	4059	12406	6.0	93	93	109	102	115	87	56	42	39	50	61	75	75
CA	Mt. Shasta	4116	12216	11.9	456	535	349	309	343	297	177	163	182	214	295	262	309
CA	Redding	4034	12224	7.9	71	86	94	81	88	89	69	62	68	68	72	70	74
CA	Montague	4144	12231	5.8	65	130	120	122	130	131	123	100	76	75	71	57	98
CA	Montague, Siskiyou Co APT	4146	12228	5.3	106	108	123	115	78	63	59	50	45	64	82	89	80
CO	La Junta	3803	10331	8.3	115	136	222	204	168	164	94	84	85	78	139	115	134
CO	Alamosa APT	3727	10552	7.4	92	110	195	254	214	167	84	70	85	91	77	74	127
CO	Pueblo, Memorial APT	3817	10431	7.7	101	122	180	231	168	129	105	84	82	81	93	104	121
CO	Colo Springs, Peterson Fld	3849	10443	9.0	142	163	217	212	189	163	99	86	105	105	138	128	142
CO	Ft. Carson, Butts AAF	3841	10446	7.3	85	93	145	218	127	131	63	71	68	112	74	87	107
CO	Denver	3945	10452	8.8	117	139	182	183	132	126	94	83	85	88	118	136	126
CO	Denver, Lowry AFB	3943	10454	8.1	115	94	131	163	112	100	95	87	102	88	126	121	109
CO	Aurura Co, Buckley Fld.	3942	10445	6.7	60	60	79	121	79	67	54	52	51	51	57	59	66
CO	Akron, Washington Co APT	4010	10313	11.7	216	313	383	359	276	239	226	184	243	212	252	280	242
CO	Rifle Co, Garfield Co. APT	3932	10744	4.1	17	32	37	69	51	39	26	23	31	25	23	15	31
CO	Craig	4031	10733	7.7	57	63	70	97	80	58	50	52	54	61	56	51	62
CT	Hartford, Bradley Fld.	4156	7241	7.7	115	127	142	129	96	75	54	53	61	74	93	100	93

				Ave. Speed Knots	Wind Power, Watts per Square Meter												
State	Location	Lat	Long		J	F	M	A	M	J	J	A	S	O	N	D	Ave.
CT	New Haven, Tweed APT	4116	7253	8.7	117	122	142	120	83	65	52	60	78	89	114	106	98
CT	Bridgeport APT	4110	7308	10.4	244	274	256	219	158	114	96	101	139	192	214	251	186
DE	Dover AFB	3908	7528	7.7	135	152	148	125	85	69	49	49	73	78	99	104	96
DE	Delaware Breakwater	3848	7506	12.7	449	570	477	430	283	196	163	190	270	403	391	410	343
DE	Wilmington, New Castle APT	3940	7536	8.1	127	149	175	147	105	85	66	59	61	84	118	126	109
DC	Washington, Andrews AFB	3848	7653	7.2	130	156	161	126	77	51	39	36	45	62	101	109	90
DC	Washinton, Bolling AFB	3850	7701	7.5	125	173	171	140	84	58	45	40	51	72	119	112	101
DC	Washington National	3851	7702	8.6	142	151	163	134	95	82	62	44	67	85	103	107	105
DC	Washington, Dulles IAP	3857	7727	6.7	104	115	118	111	66	42	37	41	40	42	66	78	68
FL	Key West NAS	2435	8147	9.5	158	172	172	176	122	98	78	71	133	133	139	147	131
FL	Homestead AFB	2529	8023	6.4	61	74	90	89	72	51	31	35	66	59	60	56	60
FL	Miami	2548	8016	7.8	87	98	111	116	80	59	58	54	90	88	78	79	80
FL	Boca Raton	2622	8006	8.2	80	108	125	135	109	72	51	55	109	140	108	106	99
FL	West Palm Beach	2643	8003	8.3	123	129	151	145	106	79	70	67	80	105	126	102	108
FL	Ft. Myers	2635	8152	7.0	93	111	153	156	104	79	58	70	99	96	90	101	101
FL	Ft. Myers, Hendricks Fld.	2638	8142	7.1	59	68	98	91	74	51	38	47	76	85	58	60	69
FL	Tampa	2758	8232	7.6	85	100	100	101	76	67	40	38	61	65	73	80	68
FL	Tamps, Macdill AFB	2751	8230	6.9	73	95	98	83	59	51	35	40	67	73	62	67	67
FL	Avon Park Range AAF	2738	8120	5.4	50	51	55	64	45	30	18	24	61	73	43	48	45
FL	Orlando, Herndon APT	2833	8120	8.2	86	110	131	120	99	83	69	85	91	107	89	99	97

MONTHLY AVERAGE WIND POWER IN THE UNITED STATES AND SOUTHERN CANADA

	MONTHLY AVERAGE WIND POWER IN THE UNITED STATES AND SOUTHERN CANADA																	
				Ave. Speed Knots	Wind Power, Watts per Square Meter												
State	Location	Lat	Long		J	F	M	A	M	J	J	A	S	O	N	D	Ave.
FL	Orlando, MCCoy AFB	2827	8118	5.9	61	76	71	67	46	41	29	24	43	48	46	51	49
FL	Titusville	2831	8047	6.7	57	72	72	58	44	42	43	32	47	56	50	56	49
FL	Cocoa Beach, Patrick AFB	2814	8036	8.8	127	149	144	134	115	80	52	62	130	191	143	127	119
FL	Cape Kennedy AFS	2829	8033	7.4	82	107	103	90	71	55	41	37	82	94	75	76	73
FL	Daytona Beach APT	2911	8103	8.9	112	141	146	142	125	94	91	95	113	161	108	116	120
FL	Jacksonville, Cecil FLD NAS	3013	8157	5.2	43	65	56	50	35	31	21	19	39	39	37	39	39
FL	Jacksonville NAS	3014	8141	6.9	61	80	81	70	58	60	40	38	76	77	62	64	63
FL	Mayport NAAS	3023	8125	7.2	82	105	92	90	67	67	40	39	110	90	74	68	76
FL	Tallahassee	3023	8422	5.8	51	59	76	66	41	28	24	28	39	43	51	51	45
FL	Marianna	3050	8511	6.9	92	104	115	86	65	48	43	36	55	61	71	84	72
FL	Panama City, Tynoall AFB	3004	8535	6.7	79	101	120	97	62	47	42	37	65	55	64	75	71
FL	Crestview	3047	8631	5.6	68	77	85	57	31	22	16	16	35	38	60	65	47
FL	Valparaiso, Eglin AFB	3029	8631	6.2	66	74	78	71	56	48	40	37	55	46	55	59	56
FL	Valparaiso, Duke Fld	3039	8632	7.0	104	123	105	115	78	46	33	38	40	48	84	88	75
FL	Valparaiso, Hurlburt Fld	3025	8641	5.5	55	62	55	51	36	31	23	21	34	33	39	45	40
FL	Milton, Whiting Fld NAAS	3042	8701	7.1	107	114	125	93	62	44	36	32	65	57	84	92	76
FL	Pensacola, Saufley Fld NAS	3026	8711	6.8	98	109	110	94	57	42	37	35	79	63	81	99	75
FL	Pensacola, Ellyson Fld	3032	8712	7.8	87	104	116	112	86	62	48	44	65	57	74	81	78
FL	Pensacola, Forest Sherman Fd	3021	8719	8.0	110	119	113	106	79	75	56	57	73	71	86	99	88
GA	Valdosta, Moody AFB	3058	8312	4.8	40	51	54	43	29	28	21	19	33	32	29	35	35

MONTHLY AVERAGE WIND POWER IN THE UNITED STATES AND SOUTHERN CANADA

State	Location	Lat	Long	Ave. Speed Knots	J	F	M	A	M	J	J	A	S	O	N	D	Ave.
GA	Moultrie	3108	8342	6.6	73	89	84	79	45	32	34	30	47	60	59	75	58
GA	Albany, Turner AFB	3135	8407	5.3	50	68	73	55	33	27	23	19	33	27	36	41	41
GA	Brunswick, Glynco NAS	3115	8128	5.5	39	54	53	52	40	36	28	25	38	39	35	37	40
GA	Ft. Stewart, Wright AAF	3153	8134	3.7	20	30	32	23	22	14	12	10	14	16	15	23	20
GA	Savannah	3208	8112	7.5	88	108	98	87	56	49	46	45	61	62	63	72	69
GA	Savannah, Hunter AFB	3201	8108	5.8	59	76	86	70	44	40	34	31	38	43	48	47	51
GA	Macon	3242	8339	8.0	92	112	103	117	69	59	56	44	61	56	68	73	75
GA	Warner Robbins AFB	3238	8336	4.9	54	76	75	59	35	26	22	18	27	31	42	44	41
GA	Ft. Benning	3221	8500	3.9	45	64	71	51	28	21	13	13	21	22	32	36	33
GA	Winder	3400	8342	7.6	99	113	93	91	57	50	51	44	43	79	92	92	78
GA	Adairsville	3455	8456	6.2	87	95	109	74	56	42	36	33	33	49	100	71	64
GA	Augusta, Bush Fld	3322	8158	5.9	68	83	87	83	43	41	36	32	43	39	45	49	53
GA	Atlanta	3339	8426	8.5	170	169	165	151	84	67	56	46	73	80	109	127	106
GA	Marietta, Dobbins AFB	3355	8432	5.8	89	99	105	96	52	38	34	30	40	50	66	72	66
HI	Honolulu IAP	2120	15055	9.8	118	131	164	163	155	172	189	194	141	128	133	144	153
HI	Barbers Point NAS	2119	15804	8.3	106	99	104	102	93	97	100	102	77	76	95	104	95
HI	Wahiawa, Wheeler AFB	2129	15802	5.9	48	49	61	59	60	70	72	65	43	40	39	49	54
HI	Waialua, Mokoleia Fld	2135	15812	7.7	59	52	97	141	115	136	151	158	113	84	89	108	109
HI	Kaneohe Bay MCAS	2127	15747	10.0	131	144	157	156	140	137	143	143	116	113	135	168	141
HI	Barking Sands AAF	2203	15947	5.6	112	62	42	40	33	24	20	22	21	38	43	69	44

	MONTHLY AVERAGE WIND POWER IN THE UNITED STATES AND SOUTHERN CANADA																		
				Ave. Speed Knots	Wind Power, Watts per Square Meter													
State	Location	Lat	Long		J	F	M	A	M	J	J	A	S	O	N	D	Ave.
HI	Molokai, Homestead Fld	2109	15706	12.3	110	195	249	291	250	312	361	342	266	268	233	238	266
HI	Kahului NAS	2054	15626	11.1	203	204	240	276	335	366	375	377	283	219	247	200	276
HI	Hilo	1943	15504	7.7	82	86	77	71	65	67	63	67	59	56	52	74	67
HI	Hilo, Lyman Fld	1943	15504	7.8	82	86	77	71	65	67	63	67	59	56	52	74	67
ID	Strevell	4201	11313	9.7	275	255	189	175	161	148	127	120	127	128	188	209	168
ID	Pocatello	4255	11236	8.6	211	224	230	209	163	159	113	87	102	103	148	176	160
ID	Idaho Falls	4331	11204	9.7	226	185	321	295	241	214	132	139	166	184	178	172	200
ID	Burley APT	4232	11346	8.0	185	162	246	199	156	116	73	58	72	89	114	150	133
ID	Twin Falls	4228	11429	8.7	168	181	232	237	155	139	85	74	86	114	131	172	147
ID	King Hill	4259	11513	8.8	220	222	357	363	330	216	169	147	212	158	165	185	221
ID	Mountain Home AFB	4303	11552	7.3	94	136	154	172	144	121	92	76	80	105	89	81	110
ID	Boise APT	4334	11613	7.8	103	112	127	119	95	79	64	56	60	76	84	95	91
IL	Chicago Midway	4147	8745	9.0	129	145	151	144	115	70	53	52	74	95	149	134	112
IL	Glenview NAS	4205	8750	8.4	164	164	203	206	137	83	56	52	72	105	143	137	128
IL	Chicago, Ohare	4159	8754	9.7	220	242	268	272	197	140	99	89	140	162	258	213	193
IL	Chicago, Ohare IAP	4159	8754	9.5	189	199	227	229	174	118	83	71	113	129	227	176	162
IL	Waterman	4146	8845	9.1	236	269	222	269	134	100	50	60	77	99	210	175	166
IL	Rockford	4212	8906	8.8	112	107	135	164	126	85	61	70	82	92	126	121	107
IL	Moline	4127	9031	8.9	121	151	215	200	155	93	63	54	91	113	185	141	130
IL	Bradford	4113	8937	10.2	210	271	284	290	203	129	68	88	96	123	237	183	196

| | | | | Ave. Speed Knots | Wind Power, Watts per Square Meter | | | | | | | | | | | | |
| | | | | | J | F | M | A | M | J | J | A | S | O | N | D | Ave. |
State	Location	Lat	Long														
IL	Rantoul, Chanute AFB	4018	8809	8.5	158	164	193	210	143	91	50	47	66	87	145	127	121
IL	Effingham	3909	8832	9.3	170	210	251	217	116	95	73	68	89	95	217	144	136
IL	Springfield, Capitol APT	3950	8940	10.6	215	253	308	295	212	131	92	82	119	152	263	242	198
IL	Quincy, Baldwin Fld	3956	9112	9.9	209	229	275	220	137	98	71	61	95	136	211	194	161
IL	Belleville, Scott AFB	3833	8951	7.2	129	140	162	143	85	61	36	33	46	76	109	96	90
IL	Marion, Williamson Co APT	3745	8901	7.6	159	186	230	243	136	88	59	45	88	93	187	28	139
IN	Evansville	3808	8732	8.1	129	139	165	154	98	69	46	38	60	71	118	117	100
IN	Terre Haute, Holman Fld	3927	8717	8.2	160	151	203	182	105	74	43	33	60	79	130	134	115
IN	Indianapolis	3944	8617	7.1	174	198	247	205	147	96	68	59	81	108	178	161	143
IN	Columbus, Bakalar AFB	3916	8554	7.0	97	106	128	117	71	50	36	32	44	58	91	88	74
IN	Milroy	3928	8522	9.5	243	270	230	209	116	115	73	67	94	101	189	163	148
IN	Centerville	3949	8458	9.0	196	237	209	182	101	87	64	57	79	93	176	136	134
IN	Marion APT	4029	8541	8.4	211	254	279	255	160	116	64	50	79	95	253	186	170
IN	Peru, Grissom AFB	4039	8609	7.7	123	137	158	165	109	65	40	36	53	69	131	133	100
IN	Lafeyette	4025	8656	10.3	290	317	290	316	175	142	91	98	112	126	296	222	215
IN	Fort Wayne	4100	8512	9.5	149	167	230	205	154	101	73	66	96	116	214	171	146
IN	Helmer	4133	8512	9.6	256	243	263	242	141	100	79	78	133	153	242	215	161
IN	Goshen	4132	8548	8.9	229	209	208	221	124	104	75	73	89	103	182	148	146
IN	South Bend	4142	8619	9.8	243	243	283	256	155	128	92	92	107	127	229	156	188
IN	McCool	4133	8710	10.7	284	297	311	290	183	149	82	96	130	157	311	231	223

MONTHLY AVERAGE WIND POWER IN THE UNITED STATES AND SOUTHERN CANADA

				Ave. Speed Knots	Wind Power, Watts per Square Meter												
State	Location	Lat	Long		J	F	M	A	M	J	J	A	S	O	N	D	Ave.
IA	Dubuque APT	4224	9042	9.4	208	209	239	317	241	150	112	135	170	198	312	231	210
IA	Burlington	4046	9107	9.5	147	160	257	164	94	85	52	44	72	97	200	144	126
IA	Iowa City APT	4138	9133	8.6	175	195	231	229	118	82	71	61	81	100	205	159	146
IA	Cedar Rapids	4153	9142	9.2	160	171	23	249	157	97	53	49	59	102	138	131	132
IA	Ottumwa	4106	9226	9.1	209	243	257	239	169	140	118	112	156	169	174	168	179
IA	Montezuma	4135	9228	11.0	270	330	330	390	256	203	103	117	150	158	271	223	237
IA	Des Moines	4132	9339	9.9	192	193	251	289	180	126	81	81	109	142	219	178	168
IA	Ft. Dodge APT	4233	9411	10.3	253	260	331	334	258	140	77	74	104	181	185	188	199
IA	Atlantic	4122	9503	11.3	296	350	363	457	295	256	136	123	155	190	264	270	256
IA	Sioux City	4224	9623	9.7	180	172	247	283	206	143	89	82	114	155	212	170	169
KS	Ft. Leavenworth	3922	9455	6.3	73	84	116	111	73	57	31	33	50	52	78	66	69
KS	Olathe NAS	3850	9453	9.2	143	157	211	187	139	117	69	69	91	102	152	136	135
KS	Topeka	3904	9538	9.8	138	147	237	229	170	159	107	112	136	137	159	163	157
KS	Topeka, Forbes AFB	3857	9540	8.6	117	134	186	185	132	115	69	79	88	95	125	104	120
KS	Ft. Riley	3903	9646	8.0	112	122	224	233	171	130	86	102	139	138	125	106	139
KS	Cassoday	3802	9638	13.0	370	436	550	550	350	310	231	257	283	284	371	311	377
KS	Wichita	3739	9725	12.0	243	273	344	337	262	276	168	177	203	221	249	237	253
KS	Wichita, McConnell AFB	3737	9716	10.9	222	234	336	317	252	237	151	136	176	188	200	207	222
KS	Hutchinson	3756	9754	10.7	287	335	372	375	330	351	215	195	309	280	308	269	305
KS	Salina, Schilling AFB	3848	9738	9.1	134	168	230	221	176	150	100	112	148	135	147	111	155

MONTHLY AVERAGE WIND POWER IN THE UNITED STATES AND SOUTHERN CANADA

MONTHLY AVERAGE WIND POWER IN THE UNITED STATES AND SOUTHERN CANADA

State	Location	Lat	Long	Ave. Speed Knots	Wind Power, Watts per Square Meter												
					J	F	M	A	M	J	J	A	S	O	N	D	Ave.
KS	Hill City APT	3923	9950	9.7	122	199	337	262	210	226	152	125	153	140	153	131	184
KS	Dodge City APT	3746	9958	13.5	281	360	441	458	360	368	259	245	296	296	334	318	336
KS	Garden City APT	3756	10043	12.4	227	326	451	450	415	456	277	272	309	259	216	204	295
KY	Corbin	3658	8408	4.4	71	54	65	58	26	15	16	12	16	18	43	44	36
KY	Lexington	3802	8436	8.9	161	158	156	169	103	75	61	47	72	73	148	146	113
KY	Warsaw	3846	8454	6.8	123	132	137	123	65	57	49	39	40	54	106	101	85
KY	Louisville, Standiford Fld	3811	8544	6.5	75	84	104	96	53	32	26	27	29	36	58	66	56
KY	Ft. Knox	3754	8558	6.6	108	121	126	111	64	46	30	25	39	47	98	96	76
KY	Bowling Green, City Co APT	3658	8626	6.6	131	115	136	113	65	39	38	32	46	57	89	93	79
KY	Ft. Campbell	3640	8730	5.8	78	88	107	89	51	32	27	25	29	39	60	69	56
KY	Paducah	3704	8846	6.7	109	106	122	106	59	44	35	33	40	48	90	93	74
LA	New Orleans	2959	9015	8.0	129	137	144	114	76	52	44	43	81	91	128	109	96
LA	New Orleans, Callender NAS	2949	9001	4.6	47	55	50	35	26	14	10	10	26	24	31	40	30
LA	Baton Rouge	3032	9109	7.4	105	106	102	95	72	52	40	36	50	53	79	92	74
LA	Lake Charles, Chenault AFB	3013	9310	8.3	184	156	204	176	125	91	58	57	67	67	133	140	122
LA	Polk AAF	3103	9311	5.7	51	69	78	68	47	37	23	15	21	29	55	51	41
LA	Alexandria, England AFB	3119	9233	4.6	45	57	64	52	37	20	15	13	17	21	39	41	35
LA	Monroe, Selman Fld	3231	9203	7.0	88	104	108	90	61	46	36	36	46	51	73	79	68
LA	Shreveport	3228	9349	8.4	128	138	145	131	92	71	57	55	58	69	105	111	97
LA	Shreveport, Barksdale AFB	3230	9340	6.0	69	74	83	72	48	36	27	27	35	34	53	59	51

					MONTHLY AVERAGE WIND POWER IN THE UNITED STATES AND SOUTHERN CANADA												
				Ave. Speed						Wind Power, Watts per Square Meter							
State	Location	Lat	Long	Knots	J	F	M	A	M	J	J	A	S	O	N	D	Ave.
ME	Portland	4339	7019	8.4	127	145	158	140	103	80	197	61	83	101	112	120	107
ME	Brunswick, NAS	4353	6956	6.8	109	116	106	107	87	64	55	48	58	69	80	97	82
ME	Bangor, Dow AFB	4448	6841	7.1	132	138	136	113	83	70	54	59	63	82	100	110	93
ME	Presque Isle AFB	4641	6803	7.8	151	167	151	161	123	88	77	69	97	115	110	134	120
ME	Limestone, Loring AFB	4657	6753	6.9	97	107	110	88	69	55	48	45	60	68	73	78	74
MD	Patuxent River NAS	3817	7625	8.1	159	177	186	148	97	76	59	59	83	102	138	139	119
MD	Baltimore, Martin Fld	3920	7625	6.9	107	111	119	95	53	44	37	39	34	46	57	64	61
MD	Baltimore, Friendship APT	3911	7640	9.6	206	253	265	209	152	117	96	79	110	117	179	188	164
MD	Ft. Mead, Tipton AAF	3905	7646	4.4	57	58	69	65	37	19	14	14	14	23	42	41	38
MD	Aberdeen, Phillips AAF	3928	7610	7.9	126	170	173	157	95	66	52	55	69	95	121	118	109
MD	Camp Detrick, Fredrick	3926	7727	5.4	101	122	144	110	51	33	25	22	30	47	96	75	72
MD	Ft. Ritchie	3944	7724	4.6	38	34	33	37	21	16	14	23	18	27	27	52	28
MA	Chicopee Falls, Westover AAF	4212	7232	7.1	122	143	131	133	96	70	52	48	60	81	104	114	96
MA	Ft. Devons AAF	4234	7136	5.4	45	40	66	84	44	31	29	32	33	39	48	53	45
MA	Bedford, Hanscom Fld	4228	7117	6.1	109	120	117	94	70	48	39	36	44	65	80	92	76
MA	Boston, Logan IAP	4222	7102	11.8	314	321	314	268	195	150	128	108	131	131	230	277	227
MA	Boston	4222	7102	11.7	314	321	314	268	195	150	128	108	131	131	230	277	227
MA	South Weymouth NAS	4209	7056	7.6	125	125	146	136	84	58	43	56	53	71	92	102	90
MA	Falmouth, Otis AFB	4139	7031	9.2	188	199	198	193	147	110	87	90	112	139	148	185	149
MA	Nantucket	4116	7003	11.6	304	346	298	277	190	140	104	113	169	214	261	298	223
MA	Nantucket Shoals	4101	6930	16.7	1024	1025	977	838	632	551	592	544	482	769	856	927	757

MONTHLY AVERAGE WIND POWER IN THE UNITED STATES AND SOUTHERN CANADA

State	Location	Lat	Long	Ave. Speed Knots	Wind Power, Watts per Square Meter												Ave.
					J	F	M	A	M	J	J	A	S	O	N	D	
MA	Georges Shoals	4141	6747	17.1	1168	1175	1058	891	619	575	519	378	473	739	891	1156	783
MI	Mt. Clemens, Selfridge AFB	4236	8249	8.2	157	151	160	145	96	71	56	53	71	84	156	144	115
MI	Ypsilanti, Willow Run	4214	8332	9.5	169	169	244	194	139	101	85	77	104	113	188	173	147
MI	Jackson	4216	8428	8.8	196	149	182	215	106	92	57	69	77	99	175	147	127
MI	Battle Creek, Kelogg APT	4218	8514	8.9	161	189	205	179	124	99	76	63	106	99	152	180	137
MI	Grand Rapids	4253	8531	8.7	120	134	180	158	112	77	62	54	83	87	162	135	113
MI	Lansing	4247	8436	10.8	273	298	356	287	178	112	69	74	123	146	251	269	203
MI	Flint, Bishop APT	4258	8344	9.6	233	206	246	195	140	109	85	71	129	140	210	223	167
MI	Saginaw, Tri City APT	4326	8352	9.7	218	196	223	196	152	111	92	74	121	128	199	189	158
MI	Muskegon Co APT	4310	8614	9.4	156	164	140	171	121	96	68	70	80	155	177	166	129
MI	Gladwin	4359	8429	5.9	67	71	92	76	63	40	31	24	34	39	61	53	53
MI	Cadillac APT	4415	8528	9.4	210	204	239	193	172	151	104	89	139	161	208	203	171
MI	Traverse City	4444	8535	9.5	229	206	249	207	147	132	100	91	160	187	250	225	178
MI	Oscoda, Wurtsmith AFB	4427	8322	7.6	117	123	121	116	91	76	55	58	71	94	109	108	94
MI	Alpena, Collins Fld	4504	8334	7.3	76	76	92	108	88	59	49	46	52	62	67	58	70
MI	Pellston, Emmett Co APT	4534	8448	8.9	183	159	192	165	154	115	105	81	115	144	175	185	147
MI	Sault Ste Marie	4628	8422	8.3	114	105	119	125	113	77	62	57	79	93	115	108	98
MI	Kinross, Kincheloe AFB	4615	8428	7.6	88	105	106	120	106	69	53	56	68	81	109	93	89
MI	Escanaba APT	4544	8705	7.8	126	164	148	186	191	150	116	93	135	163	232	143	154
MI	Gwinn, Sawyer AFB	4621	8723	7.5	94	116	109	116	100	72	52	57	65	92	102	105	90

				Ave. Speed	Wind Power, Watts per Square Meter												
State	Location	Lat	Long	Knots	J	F	M	A	M	J	J	A	S	O	N	D	Ave.
MI	Marquette	4634	8724	7.6	78	84	118	125	117	96	73	70	89	85	88	66	91
MI	Calumet	4710	8830	8.5	116	126	136	139	106	90	78	66	97	108	116	112	108
MI	Houghton Co APT	4710	8830	8.5	116	126	136	139	106	90	78	66	97	108	116	112	108
MI	Ironwood, Gogebic Co APT	4632	9008	8.5	164	198	167	290	280	174	130	139	200	213	270	203	202
MN	Minneapolis, St. Paul IAP	4453	9313	9.4	127	142	152	211	186	133	88	86	112	130	167	123	138
MN	St. Cloud, Whitney APT	4535	9411	6.9	70	70	102	129	99	71	44	38	55	66	84	58	74
MN	Alexandria	4553	9524	10.7	221	215	262	290	249	202	128	161	182	263	263	196	219
MN	Brainerd	4624	9408	6.9	90	92	111	167	134	98	62	58	107	87	123	89	102
MN	Duluth IAP	4650	9211	10.7	219	229	249	299	233	147	122	111	154	196	254	206	202
MN	Bemidji APT	4730	9456	7.2	104	120	111	244	201	155	117	120	133	141	162	122	144
MN	International Falls IAP	4834	9323	8.4	95	103	110	175	155	103	82	88	119	122	163	117	119
MN	Roseau	4851	9545	6.5	33	31	50	58	51	35	16	20	27	34	49	41	37
MN	Thief River Falls	4803	9611	8.7	215	207	194	305	279	199	135	157	189	221	309	219	219
MS	Biloxi, Keesler AFB	3024	8855	6.8	82	79	83	81	64	49	38	35	58	55	66	68	63
MS	Jackson	3220	9014	6.2	85	92	88	78	46	31	26	25	31	39	63	77	57
MS	Greenville APT	3329	9059	6.6	89	100	104	90	66	48	32	34	49	50	65	77	67
MS	Meridian NAAS	3323	8833	3.5	29	40	37	25	12	8	9	5	8	10	17	21	18
MS	Columbus AFB	3338	8827	4.7	52	60	61	48	25	17	15	13	23	21	31	40	34
MO	Malden	3636	8959	8.3	151	117	162	152	109	75	57	53	62	74	124	118	105
MO	St. Louis, Lambert Fld	3845	9023	7.9	95	116	143	138	94	60	41	36	55	60	92	96	86

MONTHLY AVERAGE WIND POWER IN THE UNITED STATES AND SOUTHERN CANADA

MONTHLY AVERAGE WIND POWER IN THE UNITED STATES AND SOUTHERN CANADA

State	Location	Lat	Long	Ave. Speed Knots	Wind Power, Watts per Square Meter												
					J	F	M	A	M	J	J	A	S	O	N	D	Ave.
MO	New Florence	3853	9126	10.1	198	231	238	231	138	105	85	85	111	112	199	165	158
MO	Kirksville	4006	9232	10.5	250	271	297	310	191	137	111	103	118	158	231	191	184
MO	Vichy, Rolla APT	3808	9146	8.6	153	158	211	170	92	72	54	45	68	76	142	156	117
MO	Ft. Leonard Wood, Forney AF	3743	9208	6.0	67	65	81	88	50	37	22	21	26	50	62	67	52
MO	Springfield	3714	9315	9.7	183	243	230	263	123	96	70	77	97	110	183	170	150
MO	Butler	3818	9420	9.3	212	208	266	226	123	124	74	65	81	131	156	160	152
MO	Knobnoster, Whiteman AFB	3844	9334	7.4	96	109	146	151	94	65	40	45	61	70	94	81	87
MO	Marshall	3906	9312	9.5	203	223	263	250	115	109	82	90	89	95	156	136	143
MO	Grandview, Rchds-Gebaur AFB	3851	9435	8.1	105	101	156	173	113	76	55	59	74	91	110	110	100
MO	Kansas City APT	3907	9436	9.4	115	126	165	182	145	129	105	99	113	112	143	122	132
MO	Knoxville	3925	9400	10.1	210	211	284	278	144	124	84	91	98	125	171	151	158
MO	Tarkio	4027	9522	8.2	121	137	258	225	192	150	87	71	85	113	135	94	135
MT	Glendive	4708	10448	7.7	131	141	145	222	217	146	125	137	139	144	111	122	149
MT	Miles City APT	4626	10552	8.6	123	137	120	161	134	102	87	98	104	109	93	116	115
MT	Wolf Point	4806	10535	8.1	117	112	143	326	248	139	126	171	245	223	183	124	179
MT	Glasgow AFB	4824	10631	8.6	133	131	125	178	198	130	102	101	138	126	119	125	133
MT	Billings, Logan Fld	4548	10832	10.0	230	210	185	202	165	137	110	99	128	152	218	237	173
MT	Livingston	4540	11032	13.5	778	819	574	415	327	239	233	253	321	500	713	1058	500
MT	Lewiston APT	4703	10927	8.6	198	185	141	163	135	108	82	95	114	125	189	153	140
MT	Havre	4834	10940	8.7	148	106	155	141	123	115	75	74	86	120	132	127	114

					Ave. Speed Knots	Wind Power, Watts per Square Meter												
State	Location	Lat	Long			J	F	M	A	M	J	J	A	S	O	N	D	Ave.
MT	Great Falls IAP	4729	11122		11.6	444	439	300	281	194	193	136	143	197	300	456	509	304
MT	Great Falls, Malmstrom AFB	4731	11110		8.9	253	240	181	176	120	112	80	80	115	178	215	263	169
MT	Helena APT	4636	11200		7.3	145	95	142	134	63	113	85	44	111	34	61	65	90
MT	Whitehall	4552	11158		11.4	710	543	352	274	221	245	193	167	174	260	410	602	344
MT	Butte, Silver Bow Co APT	4557	11230		6.9	98	101	116	158	141	120	86	85	93	93	86	76	104
MT	Missoula	4655	11405		5.0	49	36	64	75	72	70	64	41	57	23	19	26	50
NB	Omaha	4118	9554		10.0	191	186	264	280	186	148	104	104	122	158	217	191	177
NB	Omaha, Offutt AFB	4107	9555		7.6	118	123	189	198	138	98	67	58	68	93	112	112	115
NB	Grand Island APT	4058	9819		11.1	177	195	270	312	251	217	161	158	179	180	232	200	211
NB	Overton	4044	9927		10.5	209	195	323	389	276	243	155	149	162	202	299	182	222
NB	North Platte	4108	11042		10.5	193	233	374	435	321	208	153	153	206	246	234	166	254
NB	Lincoln AFB	4051	9646		9.4	163	173	258	251	193	143	97	102	102	119	173	146	162
NB	Columbus	4126	9720		10.0	184	192	316	301	246	172	113	111	120	184	150	143	186
NB	Norfolk, Stefan APT	4159	9726		9.7	236	235	308	387	275	215	142	173	204	281	361	255	256
NB	Big Springs	4105	10207		11.7	270	284	430	450	349	270	210	210	217	354	297	251	290
NB	Sidney	4108	10302		10.5	275	267	369	395	294	227	193	160	680	227	241	188	248
NB	Scottsbluff APT	4152	10336		9.8	147	225	271	254	189	180	119	124	122	166	254	199	165
NB	Alliance	4203	10248		10.6	203	209	274	358	289	233	189	204	233	228	238	210	238
NB	Valentine, Miller Fld	4252	10033		10.0	181	237	267	323	286	242	199	226	230	271	338	242	253
NV	Boulder City	3558	11450		7.6	109	162	185	230	247	293	186	197	137	95	155	81	173

MONTHLY AVERAGE WIND POWER IN THE UNITED STATES AND SOUTHERN CANADA

MONTHLY AVERAGE WIND POWER IN THE UNITED STATES AND SOUTHERN CANADA

State	Location	Lat	Long	Ave. Speed Knots	Wind Power, Watts per Square Meter												
					J	F	M	A	M	J	J	A	S	O	N	D	Ave.
NV	Las Vegas	3605	11510	8.7	105	142	186	229	225	209	166	141	111	122	67	92	150
NV	Las Vegas, Nellis AFB	3615	11502	5.7	73	89	137	138	125	123	77	75	61	63	66	57	88
NV	Indian Springs AFB	3635	11541	5.3	38	75	141	196	154	109	64	46	54	28	63	54	80
NV	Tonopah APT	3804	11708	8.7	99	150	196	196	174	142	98	94	104	113	109	101	133
NV	Fallon NAAS	3925	11843	4.7	48	50	74	72	60	49	29	23	25	30	27	36	44
NV	Reno	3930	11947	5.2	77	108	123	99	93	81	56	54	52	52	45	44	74
NV	Reno, Stead AFB	3940	11952	5.9	79	105	125	132	110	91	72	72	56	64	51	69	85
NV	Humboldt	4005	11809	6.7	61	76	140	102	103	118	98	83	66	57	47	48	79
NV	Lovelock	4004	11833	6.4	109	91	121	99	98	113	80	72	56	67	43	47	83
NV	Winnemucca APT	4054	11748	7.2	79	91	117	115	105	97	89	81	73	80	55	60	86
NV	Buffalo Valley	4020	11721	6.4	66	97	94	94	102	97	77	62	57	59	53	52	76
NV	Battle Mountain	4037	11652	7.3	148	80	147	113	132	113	85	70	61	102	64	66	98
NV	Beowawe	4036	11631	6.2	57	98	112	106	99	86	79	68	63	62	45	51	76
NV	Elko	4050	11548	6.2	68	76	100	92	99	98	91	76	75	73	52	60	76
NV	Ventosa	4052	11448	6.8	139	160	178	179	166	112	104	97	96	88	93	99	109
NH	Portsmouth, Pease AFB	4305	7049	6.6	90	112	99	82	73	50	39	37	42	54	63	90	68
NH	Manchester, Grenier Fld	4256	7126	6.7	105	150	134	137	83	68	49	37	51	72	95	127	86
NH	Keene	4254	7216	4.8	63	86	69	74	67	48	29	31	34	42	43	50	53
NJ	Atlantic City	3927	7435	9.1	185	207	207	166	109	81	62	61	81	107	144	164	129
NJ	Camden	3955	7504	8.0	132	131	167	160	86	73	64	55	62	82	118	111	104

	MONTHLY AVERAGE WIND POWER IN THE UNITED STATES AND SOUTHERN CANADA																		
				Ave. Speed Knots	Wind Power, Watts per Square Meter													
State	Location	Lat	Long		J	F	M	A	M	J	J	A	S	O	N	D	Ave.
NJ	Wrightstown, McGuire AFB	4000	7436	6.6	100	114	114	101	48	42	30	28	40	52	73	89	69
NJ	Lakehurst NAS	4002	7420	7.4	133	158	166	131	94	62	49	41	48	60	99	109	93
NJ	Belmar	4011	7404	6.1	80	82	83	57	37	31	23	23	30	50	56	55	50
NJ	Trenton	4017	7450	8.1	125	146	146	194	85	71	47	55	70	92	140	120	105
NJ	Newark	4042	7410	8.7	145	145	157	126	107	80	73	68	71	96	100	110	109
NM	Clayton	3627	10309	13.0	447	397	519	483	427	360	230	207	255	279	350	395	354
NM	Tucumcari	3511	10336	10.6	273	321	359	365	293	227	167	154	166	207	206	205	260
NM	Anton Chico	3508	10505	8.9	204	257	306	227	130	132	78	67	74	101	145	140	155
NM	Clovis, Cannon AFB	3423	10319	9.9	171	215	320	279	228	204	126	93	111	122	160	180	186
NM	Hobbs, Lea Co APT	3241	10312	10.4	195	234	353	276	250	215	138	109	118	109	166	188	190
NM	Roswell APT	3324	10432	8.5	148	191	273	260	216	172	101	82	84	102	126	172	163
NM	Roswell, Walker AFB	3318	10432	7.3	79	102	145	145	129	135	93	71	65	72	82	86	98
NM	Rodeo	3156	10859	9.4	195	216	250	325	259	191	166	131	129	155	210	173	203
NM	Las Cruces, White Sands	3222	10629	6.1	100	106	169	149	123	88	50	43	41	42	82	99	89
NM	Alamogordo, Holloman AFB	3251	10605	5.6	45	59	92	101	85	72	54	43	38	35	41	40	57
NM	Albuquerque, Kirtland AFB	3503	10637	7.6	83	115	154	190	160	134	101	72	88	97	80	74	112
NM	Otto	3505	10600	9.6	248	311	491	372	271	264	116	102	94	176	234	228	243
NM	Santa Fe APT	3537	10605	10.3	218	200	308	308	248	217	138	113	135	154	184	194	201
NM	Farmington APT	3645	10814	7.1	53	74	136	151	106	101	78	55	50	71	81	42	83
NM	Gallup	3531	10847	6.2	92	133	237	293	248	217	92	82	75	114	84	54	143

State	Location	Lat	Long	Ave. Speed Knots	J	F	M	A	M	J	J	A	S	O	N	D	Ave.
					\multicolumn{14}{c}{Wind Power, Watts per Square Meter}												
NM	Zuni	3506	10848	8.4	127	109	234	220	183	138	58	54	80	104	97	126	127
NM	El Morro	3501	10826	7.4	76	113	229	210	185	136	95	66	66	92	91	83	107
NM	Acomita	3503	10743	9.6	150	169	283	223	156	143	96	82	75	109	143	136	169
NY	Westhampton, Suffolk Co AFB	4051	7238	8.1	146	145	154	133	100	82	69	67	87	110	120	118	110
NY	Hempstead, Mitchell AFB	4044	7336	9.2	194	221	211	189	134	115	99	88	100	129	185	194	155
NY	New York, Kennedy IAP	4039	7347	10.3	242	259	260	204	151	139	122	106	120	140	173	180	168
NY	New York, La Guardia	4046	7354	10.9	300	282	283	211	160	123	105	110	135	174	217	278	197
NY	New York, Central Park	4047	7358	8.1	114	108	117	99	57	43	38	37	61	63	83	95	76
NY	New York WBO	4043	7400	11.6	436	428	384	259	211	173	146	107	143	209	329	336	261
NY	Bear Mountain	4114	7400	12.5	476	463	550	444	311	183	171	163	271	289	396	523	350
NY	Newburgh, Stewart AFB	4130	7406	7.8	164	206	193	176	108	76	61	52	64	101	136	163	124
NY	New Hackensaok	4138	7353	6.0	83	91	93	87	50	42	34	32	40	63	93	84	63
NY	Poughkeepsie, Duchess Co APT	4138	7353	6.1	68	90	106	90	52	43	33	28	36	46	66	74	60
NY	Columbiaville	4220	7345	8.7	185	220	226	173	131	104	69	69	97	138	164	172	138
NY	Albany Co APT	4245	7348	7.9	148	163	173	138	95	80	68	63	81	96	103	111	108
NY	Schnectady	4251	7357	7.4	156	116	160	155	123	81	82	67	84	68	112	114	112
NY	Plattsburg AFB	4439	7327	6.0	63	78	76	82	70	52	42	36	41	54	66	60	60
NY	Massena, Richards APT	4456	7451	9.5	176	192	217	193	150	129	108	101	111	154	170	193	158
NY	Watertown APT	4400	7601	10.0	409	312	373	298	153	134	119	99	171	197	278	350	236
NY	Rome, Griffiss AFB	4314	7525	5.7	91	108	109	94	66	44	30	26	36	50	71	82	65

MONTHLY AVERAGE WIND POWER IN THE UNITED STATES AND SOUTHERN CANADA

				Ave. Speed Knots	\multicolumn{13}{c}{Wind Power, Watts per Square Meter}												
State	Location	Lat	Long	Ave. Speed Knots	J	F	M	A	M	J	J	A	S	O	N	D	Ave.
NY	Utica, Oneida Co APT	4309	7523	8.2	117	130	113	100	74	81	43	49	59	67	106	111	87
NY	Syracuse, Hancock APT	4307	7607	8.4	158	174	166	162	108	80	66	61	76	91	129	138	115
NY	Binghampton, Bloome Co APT	4213	7559	9.0	157	183	194	191	138	77	73	70	77	104	155	160	122
NY	Elmira, Chemung Co APT	4210	7654	5.6	73	78	91	80	50	45	29	25	34	57	79	69	59
NY	Rochester	4307	7740	9.8	205	229	240	201	138	123	98	82	102	123	197	194	163
NY	Buffalo	4256	7843	11.5	468	430	417	411	422	281	278	383	377	334	354	306	382
NY	Buffalo	4256	7849	10.9	254	258	322	239	160	151	132	118	145	160	227	251	205
NY	Niagara Falls	4306	7857	8.3	193	175	147	130	106	82	72	67	82	105	134	176	126
NY	Dunkirk	4230	7916	11.2	488	348	368	302	173	154	121	127	174	269	396	361	281
NC	Wilmington	3416	7755	8.1	118	151	163	169	102	87	77	80	98	97	101	97	108
NC	Jacksonville, New Rvr. MCAF	3443	7726	6.0	59	72	84	79	53	44	32	31	43	40	49	46	51
NC	Cherry Point NAS	3454	7653	7.0	92	105	124	125	84	68	57	57	84	66	67	74	83
NC	Cape Hatteras	3516	7533	10.6	195	229	209	202	144	138	117	135	180	160	166	168	169
NC	Goldsboro, Symr-Jhnsn AFB	3520	7758	5.4	55	71	80	72	45	32	30	23	30	29	42	46	45
NC	Ft. Bragg, Simmons AAF	3508	7856	5.8	63	82	76	70	46	32	27	24	27	33	53	48	46
NC	Fayetteville, Pope AFB	3512	7901	4.3	43	54	60	55	34	25	24	21	21	23	29	30	33
NC	Charlotte, Douglas APT	3513	8056	7.4	101	101	120	118	69	57	50	53	70	76	78	82	82
NC	Asheville	3536	8232	5.5	77	77	110	90	41	24	17	16	18	33	76	74	54
NC	Hickory APT	3545	8123	7.2	69	69	89	79	57	50	50	49	49	54	63	61	62
NC	Winston Salem	3608	8014	8.1	141	166	149	169	88	68	66	54	97	106	98	117	111

MONTHLY AVERAGE WIND POWER IN THE UNITED STATES AND SOUTHERN CANADA

				Ave. Speed Knots	Wind Power, Watts per Square Meter												
State	Location	Lat	Long		J	F	M	A	M	J	J	A	S	O	N	D	Ave.
NC	Greensboro	3605	7957	6.7	67	90	94	94	47	37	34	29	35	43	69	57	58
NC	Raleigh	3552	7847	6.7	89	81	106	113	53	51	49	37	45	41	64	61	64
NC	Rocky Mount APT	3558	7748	4.2	72	74	97	86	49	62	51	43	45	50	57	60	52
NC	Elizabeth City	3616	7611	7.4	81	89	95	98	76	65	50	58	67	71	63	63	74
ND	Fargo, Hector APT	4654	9648	11.7	280	264	293	389	286	215	144	160	225	280	337	270	263
ND	Grand Forks AFB	4758	9724	8.9	167	182	183	197	166	103	71	88	123	147	147	172	146
ND	Pembina	4857	9715	11.7	308	381	321	341	335	261	187	241	261	329	403	409	308
ND	Bismarck APT	4646	10045	9.5	147	140	186	250	217	174	118	119	157	167	186	143	170
ND	Minot AFB	4825	10121	9.1	191	192	166	199	185	117	95	98	127	164	163	181	157
ND	Williston, Sloulin Fld	4811	10338	8.2	80	86	109	143	141	104	76	83	101	98	88	78	98
ND	Dickinson	4647	10248	13.0	402	365	462	486	401	402	246	208	300	332	426	334	362
OH	Youngstown APT	4116	8040	9.2	187	177	218	178	115	84	66	57	81	95	180	188	133
OH	Warren	4117	8048	9.3	196	183	197	197	116	95	67	59	82	122	164	149	136
OH	Akron	4055	8126	9.1	163	184	192	151	101	75	55	55	70	86	156	147	118
OH	Perry	4141	8107	10.7	296	296	290	277	136	115	82	90	136	108	311	270	223
OH	Cleveland	4124	8151	10.1	189	237	244	211	147	111	80	72	104	122	230	202	152
OH	Vickery	4125	8255	10.7	284	310	303	284	150	136	89	88	130	157	284	217	217
OH	Toledo	4136	8348	7.7	109	114	138	108	76	51	39	37	49	60	89	93	80
OH	Archbold	4134	8419	8.8	182	176	182	189	99	86	57	62	84	98	182	135	127
OH	Columbus	4000	8253	7.2	109	118	136	116	74	52	37	35	44	55	100	91	82

MONTHLY AVERAGE WIND POWER IN THE UNITED STATES AND SOUTHERN CANADA

				Ave. Speed Knots	Wind Power, Watts per Square Meter												
State	Location	Lat	Long		J	F	M	A	M	J	J	A	S	O	N	D	Ave.
OH	Columbus, Lockbourne AFB	3949	8256	6.8	109	127	135	120	70	53	36	33	43	58	93	94	80
OH	Hayesville	4047	8218	10.0	257	257	236	204	123	124	76	76	104	151	258	204	163
OH	Cambridge	4004	8135	6.2	110	103	110	94	54	50	40	32	40	59	30	73	71
OH	Zanesville, Cambridge	3957	8154	7.7	160	142	188	158	87	67	46	32	56	63	136	130	105
OH	Wilmington, Clinton Co AFB	3926	8348	7.8	133	148	166	157	94	60	44	37	48	62	117	119	93
OH	Cincinnati	3904	8440	8.4	133	135	150	144	90	63	51	42	61	77	126	112	99
OH	Dayton	3954	8413	9.0	179	192	207	173	108	73	59	46	69	84	170	160	125
OH	Dayton, Wright AFB	3947	8406	7.6	161	188	226	182	122	83	58	50	71	86	162	157	128
OH	Dayton, Patterson Fld	3949	8403	7.4	171	186	202	176	108	73	47	43	61	79	160	145	120
OK	Muskogee	3540	9522	8.5	149	189	230	210	132	94	53	71	80	108	114	99	130
OK	Tulsa IAP	3612	9554	9.5	157	178	196	185	155	116	94	85	110	118	145	149	141
OK	Oklahoma City	3524	9736	12.2	306	333	376	386	274	250	165	153	182	216	248	265	263
OK	Oklahoma City, Tinker AFB	3525	9723	11.4	263	277	370	412	319	318	176	153	197	230	243	248	264
OK	Ardmore AFB, Autrey Fld	3418	9701	8.7	140	165	204	203	123	113	71	72	86	98	132	114	127
OK	Ft. Sill	3439	9824	9.2	174	215	272	247	193	182	104	91	125	140	164	165	173
OK	Altus AFB	3439	9916	8.0	99	134	198	180	140	126	71	64	79	92	91	92	113
OK	Clinton-Sherman AFB	3520	9912	9.8	184	202	283	262	224	158	84	77	108	113	139	161	166
OK	Enid, Vance AFB	3620	9754	9.0	162	173	229	188	138	138	88	81	100	107	136	142	139
OK	Waynoka	3638	9850	12.4	295	416	562	556	389	310	290	250	275	309	308	261	356
OK	Gage	3618	9946	10.5	203	207	281	323	257	321	168	132	173	161	167	188	221
OR	Ontario	4401	11701	6.2	52	70	96	143	139	107	115	122	75	69	49	41	88

MONTHLY AVERAGE WIND POWER IN THE UNITED STATES AND SOUTHERN CANADA

MONTHLY AVERAGE WIND POWER IN THE UNITED STATES AND SOUTHERN CANADA

State	Location	Lat	Long	Ave. Speed Knots	Wind Power, Watts per Square Meter												
					J	F	M	A	M	J	J	A	S	O	N	D	Ave.
OR	Baker	4450	11749	7.0	44	52	47	54	47	40	40	40	39	45	38	43	46
OR	La Grande	4517	11801	8.1	328	261	173	131	93	66	55	54	60	82	201	312	152
OR	Pendleton Fld	4541	11851	8.7	96	164	196	188	174	180	134	124	134	98	127	134	145
OR	Burns	4335	11903	5.9	46	46	68	69	57	60	49	44	46	50	43	40	51
OR	Klamath Falls, Kingsley Fld	4209	12144	4.8	84	77	99	86	62	44	33	30	36	53	66	76	60
OR	Redmond, Roberts Fld	4416	12109	5.6	56	76	63	61	46	36	29	30	38	36	58	45	47
OR	Cascade Locks	4539	12150	13.1	651	718	330	331	365	351	387	353	344	451	645	750	465
OR	Crown Point	4533	12214	9.6	746	765	209	148	107	50	36	50	113	304	712	650	308
OR	Portland IAP	4536	12236	6.8	139	104	91	61	44	39	45	38	38	51	91	131	75
OR	Eugene, Mahlon Sweet Fld	4407	12313	7.6	83	86	110	94	79	74	89	74	79	58	73	77	81
OR	North Bend	4325	12413	8.4	76	128	108	107	88	185	192	149	88	52	80	94	113
OR	Roseburg	4314	12321	4.2	22	23	32	26	27	28	29	26	22	16	18	19	23
OR	Astoria, Clatsop Co APT	4609	12353	7.2	125	109	95	82	69	66	71	61	54	70	105	111	84
OR	Salem, McNary Fld	4455	12301	7.1	160	122	100	72	56	47	52	44	47	59	104	137	85
OR	Newport	4438	12404	8.5	110	109	107	95	127	145	151	111	69	72	95	129	113
OR	Wolf Creek	4241	12323	2.5	10	11	15	16	19	19	22	18	11	9	8	8	14
OR	Sexton Summit	4236	12322	11.6	323	283	243	196	236	243	255	269	248	223	316	310	276
OR	Brookings	4203	12418	6.4	91	131	96	68	58	55	35	26	37	43	72	92	63
OR	Medford	4221	12251	4.9	33	44	53	50	57	51	50	49	39	28	24	40	46
OR	Siskiyou Summit	4205	12234	8.8	96	115	102	82	129	150	170	123	102	68	89	82	109

MONTHLY AVERAGE WIND POWER IN THE UNITED STATES AND SOUTHERN CANADA

State	Location	Lat	Long	Ave. Speed Knots	Wind Power, Watts per Square Meter												
					J	F	M	A	M	J	J	A	S	O	N	D	Ave.
PA	Philadelphia	3953	7515	8.5	131	139	170	133	93	78	62	52	61	84	95	109	103
PA	Willow Grove NAS	4012	7508	6.8	111	136	148	114	74	46	35	30	43	54	88	90	81
PA	Allentown	4039	7526	7.3	157	141	227	124	77	73	38	35	55	61	108	151	104
PA	Scranton	4120	7544	7.7	87	107	94	95	80	62	47	37	50	64	85	84	74
PA	Middletown, Olmstead AFB	4012	7646	5.5	97	124	113	96	53	37	31	28	28	41	75	82	66
PA	Harrisburg	4013	7651	6.4	96	120	125	88	53	41	27	24	31	42	69	68	66
PA	Parkplace	4051	7606	12.9	464	477	451	411	251	198	132	137	211	318	411	424	345
PA	Sunbury, Selinsgrove	4053	7646	5.4	72	85	115	73	38	29	19	18	22	32	56	58	50
PA	Woodward	4055	7719	13.4	630	564	596	550	330	257	157	163	257	403	551	590	417
PA	Bellefonte	4053	7743	6.8	133	139	139	169	83	68	44	46	60	96	128	126	103
PA	Buckstown	4004	7850	9.4	261	308	321	241	131	83	73	62	83	179	235	247	192
PA	McConnellsburg	3950	7801	7.2	168	148	183	150	92	72	49	42	63	97	174	120	106
PA	Altoona, Blair Co APT	4018	7819	7.9	155	180	241	164	95	79	53	46	56	82	124	143	118
PA	Kylertown	4100	7811	9.8	268	241	302	262	147	98	85	77	98	146	202	247	200
PA	Dubois	4111	7854	7.5	118	89	127	118	79	34	36	35	41	49	77	109	78
PA	Bradford	4148	7838	6.1	81	69	76	70	49	30	21	20	25	34	54	68	49
PA	Erie IAP	4205	8011	9.1	234	191	208	147	92	77	67	62	94	111	176	216	139
PA	Mercer	4118	8012	9.0	189	169	190	176	109	80	65	58	80	107	170	163	128
PA	Brookville	4109	7906	7.4	146	119	119	132	75	60	42	45	46	75	112	104	89
PA	Pittsburg APT	4630	8013	8.5	166	170	187	162	105	75	59	50	67	82	146	150	120

				Ave. Speed Knots	Wind Power, Watts per Square Meter												
State	Location	Lat	Long		J	F	M	A	M	J	J	A	S	O	N	D	Ave.
PA	Greensburg	4016	7933	9.1	239	226	207	160	94	75	72	61	80	117	186	174	141
RI	Quonset Point NAS	4135	7125	8.4	164	164	163	156	121	84	61	70	83	115	135	141	123
RI	Providence	4144	7126	9.5	173	189	129	180	140	117	98	88	100	177	150	163	144
RI	Providence, Green APT	4144	7126	9.6	173	189	192	180	140	117	98	88	100	117	150	163	144
SC	Beaufort MCAAS	3229	8044	5.7	44	70	62	61	43	35	28	23	35	35	42	43	43
SC	Charleston	3254	8002	7.5	93	124	130	117	69	64	54	52	63	59	71	81	81
SC	Myrtle Beach AFB	3341	7856	6.1	49	65	70	78	54	51	49	44	44	40	39	40	52
SC	Florence	3411	7943	7.6	95	99	114	93	74	67	52	43	50	74	73	78	72
SC	Sumter, Shaw AFB	3358	8029	5.4	49	57	64	62	40	31	26	24	34	36	38	41	41
SC	Eastover, McIntire ANG	3355	8048	4.9	37	68	54	52	33	24	26	14	32	27	31	28	34
SC	Columbia	3357	8107	6.2	65	74	90	96	52	42	42	34	41	38	44	50	55
SC	Anderson	3430	8243	7.7	99	107	99	113	80	60	51	51	52	73	85	98	79
SC	Greenville, Donaldson AFB	3446	8223	6.3	67	70	80	79	44	38	32	27	36	37	39	50	49
SC	Spartanburg	3455	8157	8.2	121	122	156	129	95	66	65	52	60	81	108	100	94
SD	Sioux Falls, Foss Fld	4334	9644	9.5	158	156	215	266	190	129	94	91	121	144	212	147	161
SD	Watertown	4455	9709	10.1	189	224	284	332	251	233	129	123	185	210	240	151	212
SD	Aberdeen APT	4527	9826	11.2	215	234	341	413	290	244	173	177	240	249	295	202	258
SD	Huron	4423	9813	10.2	164	169	229	285	214	165	131	131	165	197	239	174	187
SD	Pierre APT	4423	10017	9.8	216	205	255	294	202	141	124	129	149	168	230	207	191
SD	Rapid City	4403	10304	9.6	176	173	239	234	178	144	123	139	168	193	285	205	191

MONTHLY AVERAGE WIND POWER IN THE UNITED STATES AND SOUTHERN CANADA

MONTHLY AVERAGE WIND POWER IN THE UNITED STATES AND SOUTHERN CANADA																	
				Ave. Speed	Wind Power, Watts per Square Meter												
State	Location	Lat	Long	Knots	J	F	M	A	M	J	J	A	S	O	N	D	Ave.
SD	Rapid City, Ellsworth AFB	4409	10306	9.9	306	256	382	354	245	191	164	172	196	233	320	294	260
SD	Hot Springs	4322	10323	8.2	90	117	155	297	219	163	94	128	126	136	179	125	153
TN	Bristol	3630	8221	5.9	76	105	93	85	53	40	30	30	22	37	51	59	60
TN	Knoxville APT	3549	8359	7.0	133	135	156	161	85	66	53	42	47	54	99	100	94
TN	Chattanooga	3502	8512	5.6	64	70	76	83	42	31	26	20	26	31	49	52	48
TN	Chattanooga	3503	8512	5.4	70	86	106	83	46	44	37	29	35	42	67	60	60
TN	Monteagle	3515	8550	5.4	77	84	84	63	32	19	18	16	19	33	60	70	48
TN	Smyrna, Sewart AFB	3600	8632	5.2	76	83	87	82	42	29	22	20	22	31	58	62	51
TN	Nashville, Berry Fld	3607	8641	7.4	116	114	132	117	70	67	41	33	44	57	87	80	80
TN	Memphis NAS	3521	8952	6.2	84	85	93	84	54	37	25	24	29	36	67	73	58
TN	Memphis IAP	3503	8959	7.9	130	137	148	125	89	57	45	42	54	61	102	110	89
TX	Brownsville, Rio Grande IAP	2554	9726	10.7	231	229	281	292	277	211	185	141	102	104	150	186	199
TX	Harlington AFB	2614	9740	8.8	124	175	212	207	172	160	132	129	82	75	101	109	139
TX	Kingsville NAAS	2731	9749	8.5	111	130	162	183	171	157	142	115	107	77	104	98	129
TX	Corpus Christi	2746	9730	10.4	189	229	260	252	199	177	158	150	107	114	157	154	179
TX	Corpus Christi NAS	2742	9716	11.3	209	232	272	286	263	225	189	153	150	144	206	172	210
TX	Laredo AFB	2732	9928	10.0	90	122	153	185	206	223	216	174	122	101	93	82	147
TX	Beeville NAAS	2823	9740	7.3	81	101	124	129	111	85	71	61	60	52	76	72	85
TX	Victoria, Foster AFB	2851	9655	7.9	134	173	198	138	112	97	67	71	53	58	103	124	109
TX	Houston	2939	9517	10.1	191	216	240	258	186	139	85	74	101	117	185	159	163

MONTHLY AVERAGE WIND POWER IN THE UNITED STATES AND SOUTHERN CANADA

State	Location	Lat	Long	Ave. Speed Knots	Wind Power, Watts per Square Meter												
					J	F	M	A	M	J	J	A	S	O	N	D	Ave.
TX	Houston, Ellington AFB	2937	9510	6.8	86	96	114	104	81	56	38	41	56	54	82	75	72
TX	Galveston AAF	2916	9451	11.0	262	261	298	257	227	200	149	144	147	148	239	217	210
TX	Pt. Arthur, Jefferson Co APT	2957	9401	9.3	153	186	184	190	156	95	64	54	115	85	121	134	128
TX	Lufkin, Angelina Co APT	3114	9445	6.1	65	72	76	70	45	30	27	22	26	38	56	62	49
TX	Saltillo	3312	9519	8.7	131	172	171	185	100	86	72	58	71	84	102	104	112
TX	San Antonio	2932	9028	8.1	99	113	114	123	104	104	83	62	65	71	94	87	98
TX	San Antonio, Randolph AFB	2932	9817	7.3	94	105	115	109	94	82	64	58	61	56	90	82	84
TX	San Antonio, Kelly AFB	2923	9835	6.9	84	85	105	109	96	83	63	53	52	53	72	63	77
TX	San Antonio, Brooks AFB	2921	9827	8.9	141	144	185	189	187	166	136	108	95	107	135	103	142
TX	Hondo AAF	2920	9910	6.7	59	82	88	94	99	96	51	45	42	32	50	49	64
TX	Kerrville	2959	9905	7.1	86	95	134	129	117	92	97	48	52	63	74	54	86
TX	San Marcos	2953	9752	7.2	115	121	142	115	127	102	64	65	56	78	108	105	98
TX	Austin, Bergstrom AFB	3012	9740	7.8	145	140	167	145	118	119	88	73	58	74	117	116	115
TX	Bryan	3038	9628	7.0	85	104	108	103	89	74	49	48	38	47	73	89	76
TX	Killeen, Fort Hood AAF	3108	9743	8.1	123	138	151	155	130	109	81	59	60	75	96	114	106
TX	Ft Hood, Gray AAF	3104	9750	9.2	163	179	198	208	162	153	121	87	72	102	147	158	146
TX	Waco, Connally AFB	3138	9704	7.7	117	111	135	128	101	90	72	61	56	68	104	101	95
TX	Dallas NAS	3244	9658	9.1	155	166	210	199	154	144	97	82	85	99	135	131	137
TX	Ft. Worth, Carswell AFB	3246	9725	8.2	138	154	216	194	141	133	71	60	72	89	129	121	124
TX	Mineral Wells APT	3247	9804	9.2	120	146	203	201	162	154	103	79	78	88	111	110	128

| | | | | | MONTHLY AVERAGE WIND POWER IN THE UNITED STATES AND SOUTHERN CANADA | | | | | | | | | | | | |

State	Location	Lat	Long	Ave. Speed Knots	Wind Power, Watts per Square Meter												
					J	F	M	A	M	J	J	A	S	O	N	D	Ave.
TX	Mineral Wells, Ft Walters AAF	3250	9803	8.7	117	135	192	184	149	136	92	71	71	82	99	103	11
TX	Santo	3237	9814	7.1	92	136	157	163	87	83	55	55	60	55	112	77	9
TX	Sherman, Perrin AFB	3343	9640	9.1	172	164	217	213	142	121	79	73	81	106	153	153	13
TX	Gainsville	3340	9708	11.1	244	303	338	357	226	184	141	134	155	163	222	188	22
TX	Wichita Falls	3358	9829	9.9	160	178	244	222	176	158	110	95	107	114	168	154	15
TX	Abilene, Dyers AFB	3226	9951	7.7	91	104	145	147	124	102	61	51	58	66	87	87	9
TX	San Angelo, Mathis Fld	3122	10030	8.9	117	154	195	184	170	147	96	86	90	90	113	108	12
TX	San Angelo, Goodfellow AFB	3124	10024	8.7	108	152	191	179	169	157	85	81	91	88	114	102	12
TX	Del Rio, Laughlin AFB	2922	10045	7.6	71	107	114	120	118	117	91	68	59	57	56	60	8
TX	Canadian	3500	10022	13.0	390	416	570	596	430	344	263	231	310	337	370	290	37
TX	Dalhart APT	3601	10233	12.9	370	371	476	477	477	559	334	278	280	228	266	305	35
TX	Amarillo, English Fld	3514	10142	11.7	229	279	359	329	290	240	166	135	180	200	221	222	24
TX	Childress	3426	10017	10.2	152	194	272	269	221	204	118	93	117	126	128	147	17
TX	Lubbock, Reese AFB	3336	10203	9.4	155	211	291	268	204	188	89	67	88	99	140	169	16
TX	Big Spring, Webb AFB	3213	10131	10.1	155	197	264	266	226	216	128	101	112	124	135	139	17
TX	Midland	3156	10212	8.9	91	143	146	155	133	123	94	76	85	84	89	102	10
TX	Wink, Winkler Co APT	3147	10312	8.5	95	148	204	181	181	193	119	78	72	75	81	128	11
TX	Marfa APT	3016	10401	7.9	128	182	165	192	146	117	84	65	85	84	88	119	12
TX	Guadalupe Pass	3150	10448	15.8	887	892	999	932	868	603	422	342	401	555	760	827	71
TX	El Paso	3148	10629	9.8	176	257	296	299	222	173	131	111	102	128	153	163	18

				Ave. Speed Knots	Wind Power, Watts per Square Meter												
State	Location	Lat	Long		J	F	M	A	M	J	J	A	S	O	N	D	Ave.
TX	El Paso, Biggs AFB	3150	10624	5.8	68	99	144	144	96	74	47	38	32	33	49	59	72
UT	St George	3703	11331	5.1	26	39	73	67	72	74	53	62	40	25	34	23	49
UT	Milford	3826	11300	10.2	179	241	228	275	302	241	220	175	167	173	172	152	214
UT	Bryce Canyon APT	3742	11209	6.4	69	65	93	79	94	90	40	47	53	48	53	49	66
UT	Hanksville	3822	11043	4.6	43	41	115	84	102	108	35	38	43	40	36	25	57
UT	Tooele, Dugway PG	4011	11256	4.8	38	48	69	81	71	69	52	57	46	41	31	28	53
UT	Darby, Wendover AFB	4043	11402	5.3	59	62	90	90	71	82	62	61	49	50	58	36	62
UT	Wendover	4044	11402	5.4	48	55	91	103	84	80	57	57	43	43	47	40	62
UT	Locomotive Springs	4143	11255	9.3	113	129	205	193	221	215	195	202	167	125	115	93	185
UT	Ogden, Hill AFB	4107	11158	8.0	102	118	127	126	130	124	124	130	122	123	99	92	119
UT	Salt Lake City	4046	11158	7.7	77	84	100	100	96	96	82	106	73	68	66	71	85
UT	Coalville	4054	11125	3.9	26	35	39	32	30	28	17	18	37	24	17	24	28
VT	Montpelier, Barre APT	4412	7234	7.2	162	167	152	118	99	97	69	60	94	103	102	125	111
VT	Burlington, Ethan Allen AB	4428	7309	7.7	114	111	103	100	85	70	55	52	70	82	100	114	90
VA	Norfolk NAS	3656	7618	8.8	154	182	171	139	103	84	75	80	111	127	130	128	121
VA	Oceana NAS	3650	7601	7.6	135	136	150	126	84	62	52	52	83	93	97	107	98
VA	Hampton, Langley AFB	3705	7622	8.5	156	187	193	166	122	86	73	81	117	137	135	146	136
VA	Ft. Eustis, Felker AAF	3708	7636	6.5	79	90	89	74	51	42	32	32	43	46	61	64	58
VA	South Boston	3641	7855	5.0	49	49	65	65	34	33	32	24	28	36	40	38	40
VA	Danville APT	3634	7920	6.1	69	63	84	78	43	36	32	31	33	35	41	39	48

MONTHLY AVERAGE WIND POWER IN THE UNITED STATES AND SOUTHERN CANADA

	MONTHLY AVERAGE WIND POWER IN THE UNITED STATES AND SOUTHERN CANADA																
				Ave. Speed Knots	Wind Power, Watts per Square Meter												
State	Location	Lat	Long		J	F	M	A	M	J	J	A	S	O	N	D	Ave.
VA	Roanoke	3719	7958	7.1	152	207	171	144	75	55	51	46	46	53	98	132	102
VA	Richmond	3730	7720	6.7	59	68	79	74	49	40	34	32	39	40	47	47	51
VA	Quantico MCAS	3830	7719	6.0	55	63	75	67	45	35	29	29	35	35	46	45	47
VA	Ft.Belvoir, Davison AAF	3843	7711	3.8	42	60	60	42	24	15	12	12	13	19	34	44	31
WA	Spokane IAP	4738	11732	7.2	92	111	106	102	73	68	55	53	58	63	80	95	79
WA	Spokane, Fairchild AFB	4738	11739	7.2	118	138	134	121	92	87	67	62	74	84	92	123	100
WA	Moses Lake, Larson AFB	4711	11919	6.2	68	62	98	101	79	80	56	45	61	55	53	55	68
WA	Walla Walla	4606	11817	6.7	87	100	114	93	68	65	56	55	51	47	86	89	76
WA	Pasco, Tri City APT	4616	11907	6.8	342	331	346	372	243	212	190	224	229	186	247	164	257
WA	North Dalles	4537	12109	8.0	65	72	171	189	264	279	334	284	166	93	63	64	170
WA	Yakima	4634	12032	6.4	60	53	89	115	78	71	54	48	52	48	41	40	62
WA	Chehalis	4640	12205	6.4	109	86	82	59	53	44	45	44	43	57	80	110	68
WA	Kelso, Castle Rock	4608	12254	6.9	128	107	87	65	65	46	46	39	52	66	119	133	80
WA	North Head	4616	12404	13.0	547	521	495	369	430	357	330	275	215	366	460	773	428
WA	Hoquium, Bowerman APT	4658	12356	8.2	141	112	108	88	86	66	59	56	53	91	94	103	88
WA	Moclips	4715	12412	7.6	82	82	75	81	68	38	37	36	41	65	69	102	65
WA	Tatoosh IS	4823	12444	12.3	763	603	443	269	276	117	130	109	195	422	569	735	388
WA	Tacoma, McChord AFB	4709	12229	4.6	50	48	53	49	38	30	25	23	25	31	41	41	38
WA	Ft. Lewis, Gray AAF	4705	12235	3.9	36	28	30	30	23	19	17	18	17	21	23	27	24
WA	Seattle Tacoma	4727	12218	9.5	194	210	211	173	130	120	94	84	104	135	147	195	149
WA	Seattle FWC	4741	12216	5.6	69	61	58	48	32	28	25	24	28	44	53	69	45

MONTHLY AVERAGE WIND POWER IN THE UNITED STATES AND SOUTHERN CANADA

State	Location	Lat	Long	Ave. Speed Knots	Wind Power, Watts per Square Meter												
					J	F	M	A	M	J	J	A	S	O	N	D	Ave.
WA	Everett, Paine AFB	4755	12217	6.3	70	67	66	57	44	40	38	34	39	43	60	65	51
WA	Whidbey IS NAS	4821	12240	7.1	185	158	148	124	75	54	43	33	50	104	160	188	108
WA	Bellingham APT	4848	12232	6.3	131	132	92	66	42	43	43	35	26	56	89	119	71
WV	Charleston	3822	8136	5.6	46	61	62	53	37	28	24	15	22	21	45	45	38
WV	Elkins, Randolph Co APT	3853	7951	5.8	80	90	101	92	57	32	24	22	25	39	71	68	58
WV	Morgantown APT	3939	7955	5.8	73	73	86	67	37	26	18	16	22	34	64	81	48
WI	Green Bay	4429	8808	5.6	174	150	212	197	173	129	89	72	122	131	192	151	149
WI	Milwaukee, Mitchell Fld	4257	8754	10.2	198	212	246	238	195	120	95	91	129	159	229	200	175
WI	Madison, Traux Fld	4308	8920	8.8	139	148	199	197	153	97	72	62	93	112	170	136	130
WI	Camp Douglas, Volk Fld	4356	9016	6.3	62	76	79	77	64	33	28	26	34	64	76	56	54
WI	La Crosse APT	4352	9115	8.8	120	116	145	201	171	100	70	69	100	128	179	134	127
WI	Eau Claire	4452	9129	8.3	96	113	102	168	155	90	83	85	100	109	134	106	112
WY	Cheyenne APT	4109	10449	11.9	433	453	434	399	242	176	125	132	157	220	402	463	302
WY	Laramie	4118	10540	11.5	498	506	520	339	313	300	152	173	212	259	338	379	312
WY	Medicine Bow	4153	10611	12.7	773	758	825	518	343	296	250	223	328	423	544	726	490
WY	Bitter Creek	4140	10833	12.7	477	550	623	397	297	230	150	223	210	284	370	477	357
WY	Rock Springs	4138	10915	10.6	503	476	551	357	311	264	216	189	236	250	320	394	339
WY	Casper AAF	4255	10627	11.4	445	417	359	266	201	222	145	150	219	208	362	473	289
WY	Sheridan	4446	10658	6.6	71	71	80	94	88	73	60	56	62	65	76	64	72
NS	Yarmouth	4350	6605	9.0	230	173	190	153	109	84	63	67	84	123	162	196	136

				Ave. Speed Knots	Wind Power, Watts per Square Meter												
State	Location	Lat	Long		J	F	M	A	M	J	J	A	S	O	N	D	Ave.
NB	Frederickton	4552	6632	7.7	153	124	148	116	99	86	69	64	70	94	98	127	104
QU	Mont Jolt	4836	6812	11.2	356	358	310	220	197	157	134	152	188	245	297	381	250
QU	Bagotville	4820	7100	9.3	206	180	202	166	164	141	95	97	128	147	181	148	155
QU	St. Hubert	4531	7325	9.3	224	213	163	153	145	138	100	83	110	136	203	177	154
ON	Ottawa	4519	7540	8.2	123	120	125	111	99	79	58	53	70	88	117	100	95
ON	Trenton	4407	7732	8.9	203	176	166	148	125	103	94	80	100	122	181	151	137
ON	Toronto	4341	7938	8.6	198	177	157	150	110	87	69	68	85	105	177	152	130
ON	London	4302	8109	9.2	239	229	233	214	139	85	66	65	82	107	173	176	151
ON	Sudbury	4637	8048	12.2	318	392	330	324	328	290	218	195	252	294	355	312	301
ON	White River	4836	8517	4.3	19	26	28	32	37	33	24	21	24	28	30	24	27
ON	Kenora	4948	9422	8.6	91	95	91	111	104	74	63	70	88	96	112	86	90
MN	Winnipeg	4954	9714	10.7	206	218	227	293	266	177	116	138	174	209	239	210	206
MN	Rivers	5001	10019	10.5	216	171	187	266	276	194	137	149	200	235	218	200	204
SA	Regina	5026	10440	11.9	327	286	315	350	368	243	162	186	286	232	279	300	278
SA	Moose Jaw	5023	10534	12.3	399	343	314	344	390	299	197	214	340	319	340	367	322
AL	Medicine Hat	5001	11043	8.9	164	159	146	215	176	138	96	110	159	168	187	177	158
AL	Lethbridge	4938	11248	12.5	625	563	356	450	370	319	202	246	279	510	546	567	419
BC	Penticton	4928	11936	7.5	265	187	141	104	76	64	52	49	64	132	233	274	137
BC	Vancouver	4911	12310	6.5	72	75	85	83	53	48	52	39	48	62	80	75	64
BC	Victoria	4839	12326	6.5	84	80	74	75	50	51	34	36	36	47	68	79	60

MONTHLY AVERAGE WIND POWER IN THE UNITED STATES AND SOUTHERN CANADA

Wind Power Maps

This appendix presents five maps indicating roughly the wind power available near ground level in the continental United States and southern Canada. They are derived from the same report as the data in Appendix 2.3, *Wind Power Climatology in the United States*, by Jack Reed. The first map shows the *annual* average wind power available, and the other four indicate seasonal average wind power for spring, summer, fall and winter.

Values of wind power are given in units of watts per square meter along *isodyns*, or curves of constant power. To convert to watts per square foot, multiply by 0.0929. For comparison, note that windspeeds of 10, 15 and 20 mph yield wind power values of 55, 185 and 438 watts per square meter at sea level. These maps indicate only the total wind power available in the wind; the actual wind power extracted by a particular machine will be some fraction, usually 10–50 percent of that available.

The windspeed measurements upon which these maps are based were made at differing anemometer height, even though the standard height is 10 meters (33 feet). No attempt was made to correct for these differences or for the effects of local terrain features on wind patterns. Small variations in windspeed from one station to the next are amplified by the cubic dependence of wind power on windspeed. These measurements were smoothed by averaging all values over circular areas 300 miles in diameter before making the maps. The isodyns were also adjusted subjectively to conform to major geographic features like mountain ranges and coastlines.

These maps indicate considerable wind power is available in the Western Great Plains and along the New England and Pacific Northwest coastlines. Southeastern and southwestern areas of the United States have very limited wind power potential.

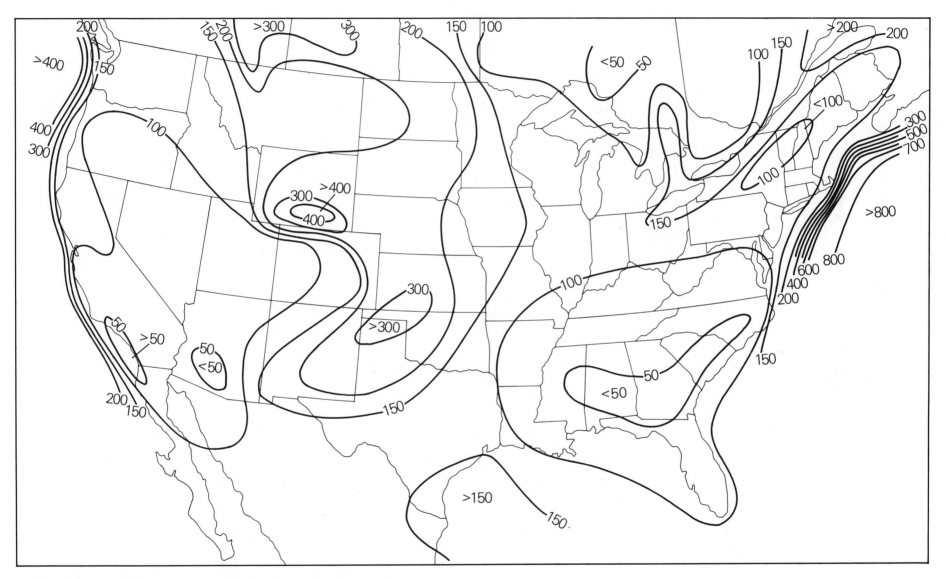

Annual Average Wind Power Available, in watts per square meter.

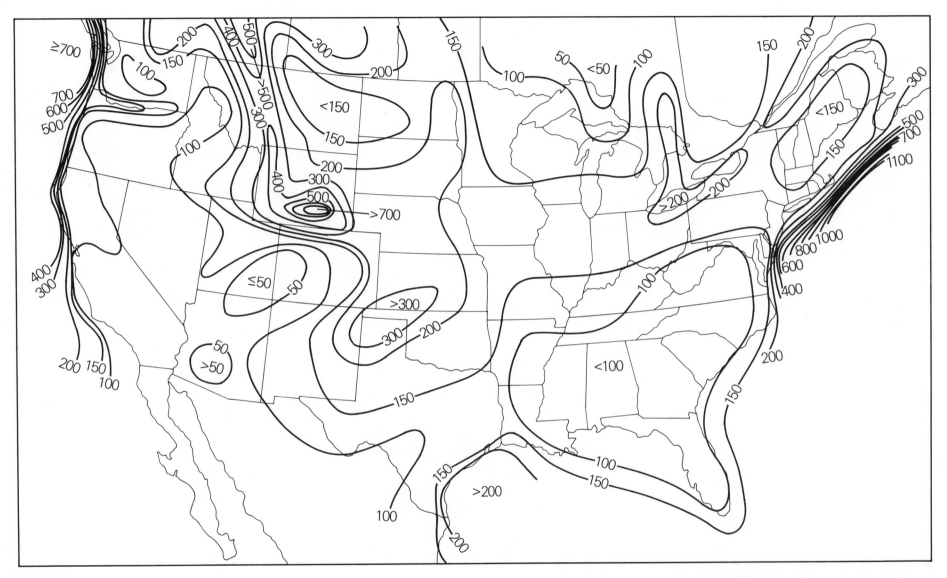

Winter Average Wind Power Available, in watts per square meter.

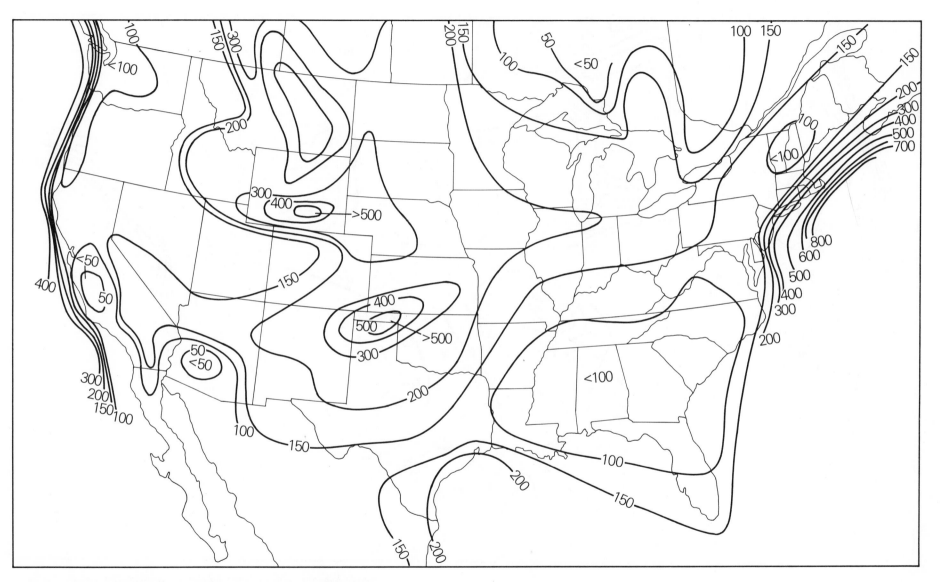

Spring Average Wind Power Available, in watts per square meter.

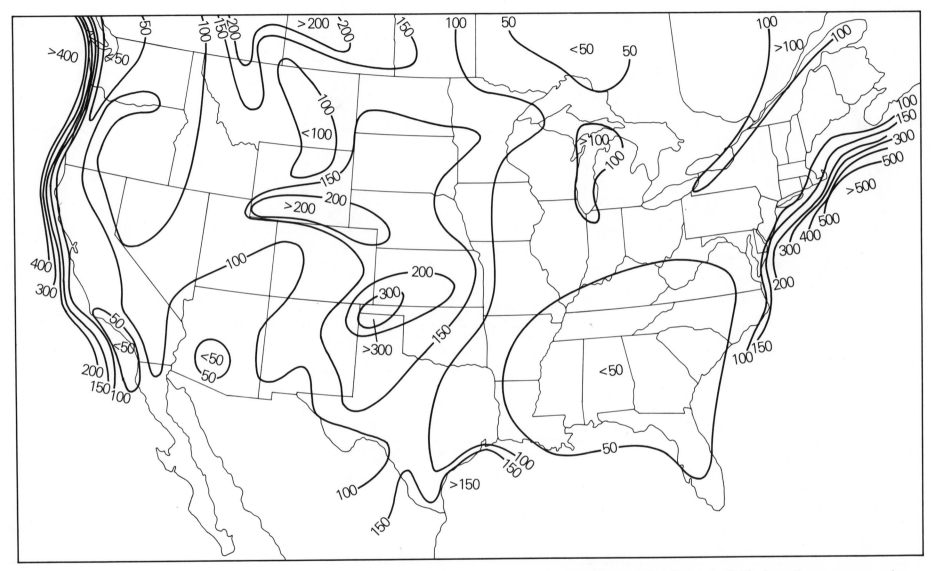

Summer Average Wind Power Available, in watts per square meter.

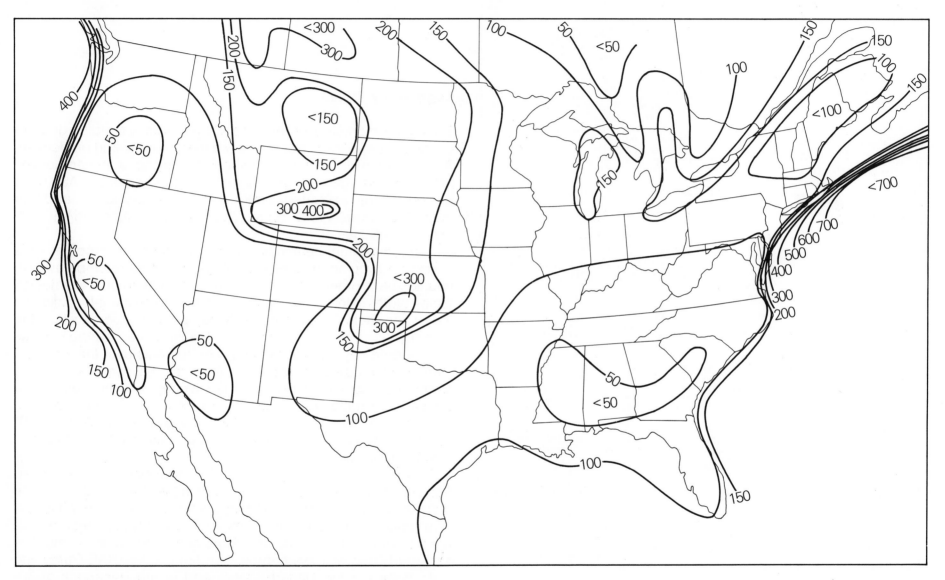

Autumn Average Wind Power Available, in watts per square meter.

State	City	Avg. Winter Temp	Design Temp	Sep	Oct	Nov	Dec	Jan	Feb	Mar	Apr	May	Yearly Total
Ala.	Birmingham	54.2	19	6	93	363	555	592	462	363	108	9	2551
	Huntsville	51.3	13	12	127	426	663	694	557	434	138	19	3070
	Mobile	59.9	26	0	22	213	357	415	300	211	42	0	1560
	Montgomery	55.4	22	0	68	330	527	543	417	316	90	0	2291
Alaska	Anchorage	23.0	−25	516	930	1284	1572	1631	1316	1293	879	592	10864
	Fairbanks	6.7	−53	642	1203	1833	2254	2359	1901	1739	1068	555	14279
	Juneau	32.1	− 7	483	725	921	1135	1237	1070	1073	810	601	9075
	Nome	13.1	−32	693	1094	1455	1820	1879	1666	1770	1314	930	14171
Ariz.	Flagstaff	35.6	0	201	558	867	1073	1169	991	911	651	437	7152
	Phoenix	58.5	31	0	22	234	415	474	328	217	75	0	1765
	Tucson	58.1	29	0	25	231	406	471	344	242	75	6	1800
	Winslow	43.0	9	6	245	711	1008	1054	770	601	291	96	4782
	Yuma	64.2	37	0	0	108	264	307	190	90	15	0	974
Ark.	Fort Smith	50.3	9	12	127	450	704	781	596	456	144	22	3292
	Little Rock	50.5	19	9	127	465	716	756	577	434	126	9	3219
	Texarkana	54.2	22	0	78	345	561	626	468	350	105	0	2533
Calif.	Bakersfield	55.4	31	0	37	282	502	546	364	267	105	19	2122
	Burbank	58.6	36	6	43	177	301	366	277	239	138	81	1646
	Eureka	49.9	32	258	329	414	499	546	470	505	438	372	4643
	Fresno	53.3	28	0	84	354	577	605	426	335	162	62	2611
	Long Beach	57.8	36	9	47	171	316	397	311	264	171	93	1803
	Los Angeles	57.4	41	42	78	180	291	372	302	288	219	158	2061
	Oakland	53.5	35	45	127	309	481	527	400	353	255	180	2870
	Sacramento	53.9	30	0	56	321	546	583	414	332	178	72	2502
	San Diego	59.5	42	21	43	135	236	298	235	214	135	90	1458
	San Francisco	55.1	42	102	118	231	388	443	336	319	279	239	3001
	Santa Maria	54.3	32	96	146	270	391	459	370	363	282	233	2967
Colo.	Alamosa	29.7	−17	279	639	1065	1420	1476	1162	1020	696	440	8529
	Colorado Springs	37.3	− 1	132	456	825	1032	1128	938	893	582	319	6423
	Denver	37.6	− 2	117	428	819	1035	1132	938	887	558	288	6283
	Grand Junction	39.3	8	30	313	786	1113	1209	907	729	387	146	5641
	Pueblo	40.4	− 5	54	326	750	986	1085	871	772	429	174	5462
Conn.	Bridgeport	39.9	4	66	307	615	986	1079	966	853	510	208	5617
	Hartford	37.3	1	117	394	714	1101	1190	1042	908	519	205	6235
	New Haven	39.0	5	87	347	648	1011	1097	991	871	543	245	5897
Del.	Wilmington	42.5	12	51	270	588	927	980	874	735	387	112	4930
D. C.	Washington	45.7	16	33	217	519	834	871	762	626	288	74	4224
Fla.	Daytona Beach	64.5	32	0	0	75	211	248	190	140	15	0	879
	Fort Myers	68.6	38	0	0	24	109	146	101	62	0	0	442
	Jacksonville	61.9	29	0	12	144	310	332	246	174	21	0	1239
	Key West	73.1	55	0	0	0	28	40	31	9	0	0	108
	Lakeland	66.7	35	0	0	57	164	195	146	99	0	0	661
	Miami	71.1	44	0	0	0	65	74	56	19	0	0	214
	Miami Beach	72.5	45	0	0	0	40	56	36	9	0	0	141
	Orlando	65.7	33	0	0	72	198	220	165	105	6	0	766
	Pensacola	60.4	29	0	19	195	353	400	277	183	36	0	1463
	Tallahassee	60.1	25	0	28	198	360	375	286	202	36	0	1485

Degree-Day Tables

The hourly, monthly and yearly heat losses from a building depend on the temperature difference between the indoor and outdoor air. To aid in the estimation of these heat losses, the American Society of Heating, Refrigeration and Air-conditioning Engineers (ASHRAE) publishes the expected winter design temperatures and the monthly and yearly total degree-days for many cities and towns in the United States.

The maximum heat loss rate occurs when the temperature is lowest, and you need some idea of the lowest likely temperature in your locale in order to size a conventional heating unit. The design temperatures in the accompanying tables provide this information for the United States. They indicate the temperatures below which the mercury falls for only 2½ percent of the time in an average winter.

Degree-days gauge heating requirements over the long run. One degree-day accrues for every day the average outdoor temperature is 1°F below 65°F. For example, if the outdoor air temperature remained constant at 30°F for the entire month of January, then $31 \times (65 - 30) = 1085$ degree-days would result. Of course, the temperature varies over the course of a day, so this is an oversimplified example.

Both monthly and yearly total degree-days for the United States and southern Canada are listed in these tables, but only the months of September through May are included here because very little heating is needed in the summer. The yearly total degree-days are the sum over *all 12 months*. More complete listings of monthly and yearly degree-days can be found in the ASHRAE *Guide and Data Book* (for the United States), and in the *Handbook of Air-Conditioning, Heating and Ventilating* (for Canada).

State	City	Avg. Winter Temp	Design Temp	Sep	Oct	Nov	Dec	Jan	Feb	Mar	Apr	May	Yearly Total
	Tampa	66.4	36	0	0	60	171	202	148	102	0	0	683
	West Palm Beach	68.4	40	0	0	6	65	87	64	31	0	0	253
Ga.	Athens	51.8	17	12	115	405	632	642	529	431	141	22	2929
	Atlanta	51.7	18	18	124	417	648	636	518	428	147	25	2961
	Augusta	54.5	20	0	78	333	552	549	445	350	90	0	2397
	Columbus	54.8	23	0	87	333	543	552	434	338	96	0	2383
	Macon	56.2	23	0	71	297	502	505	403	295	63	0	2136
	Rome	49.9	16	24	161	474	701	710	577	468	177	34	3326
	Savannah	57.8	24	0	47	246	437	437	353	254	45	0	1819
Hawaii	Hilo	71.9	59	0	0	0	0	0	0	0	0	0	0
	Honolulu	74.2	60	0	0	0	0	0	0	0	0	0	0
Idaho	Boise	39.7	4	132	415	792	1017	1113	854	722	438	245	5809
	Lewiston	41.0	6	123	403	756	933	1063	815	694	426	239	5542
	Pocatello	34.8	−8	172	493	900	1166	1324	1058	905	555	319	7033
Ill.	Chicago	37.5	−4	81	326	753	1113	1209	1044	890	480	211	6155
	Moline	36.4	−7	99	335	774	1181	1314	1100	918	450	189	6408
	Peoria	38.1	−2	87	326	759	1113	1218	1025	849	426	183	6025
	Rockford	34.8	−7	114	400	837	1221	1333	1137	961	516	236	6830
	Springfield	40.6	−1	72	291	696	1023	1135	935	769	354	136	5429
Ind.	Evansville	45.0	6	66	220	606	896	955	767	620	237	68	4435
	Fort Wayne	37.3	0	105	378	783	1135	1178	1028	890	471	189	6205
	Indianapolis	39.6	0	90	316	723	1051	1113	949	809	432	177	5699
	South Bend	36.6	−2	111	372	777	1125	1221	1070	933	525	239	6439
Iowa	Burlington	37.6	−4	93	322	768	1135	1259	1042	859	426	177	6114
	Des Moines	35.5	−7	96	363	828	1225	1370	1137	915	438	180	6588
	Dubuque	32.7	−11	156	450	906	1287	1420	1204	1026	546	260	7376
	Sioux City	34.0	−10	108	369	867	1240	1435	1198	989	483	214	6951
	Waterloo	32.6	−12	138	428	909	1296	1460	1221	1023	531	229	7320
Kans.	Dodge City	42.5	3	33	251	666	939	1051	840	719	354	124	4986
	Goodland	37.8	−2	81	381	810	1073	1166	955	884	507	236	6141
	Topeka	41.7	3	57	270	672	980	1122	893	722	330	124	5182
	Wichita	44.2	5	33	229	618	905	1023	804	645	270	87	4620
Ky.	Covington	41.4	3	75	291	669	983	1035	893	756	390	149	5265
	Lexington	43.8	6	54	239	609	902	946	818	685	325	105	4683
	Louisville	44.0	8	54	248	609	890	930	818	682	315	105	4660
La.	Alexandria	57.5	25	0	56	273	431	471	361	260	69	0	1921
	Baton Rouge	59.8	25	0	31	216	369	409	294	208	33	0	1560
	Lake Charles	60.5	29	0	19	210	341	381	274	195	39	0	1459
	New Orleans	61.0	32	0	19	192	322	363	258	192	39	0	1385
	Shreveport	56.2	22	0	47	297	477	552	426	304	81	0	2184
Me.	Caribou	24.4	−18	336	682	1044	1535	1690	1470	1308	858	468	9767
	Portland	33.0	−5	195	508	807	1215	1339	1182	1042	675	372	7511
Md.	Baltimore	43.7	12	48	264	585	905	936	820	679	327	90	4654
	Frederick	42.0	7	66	307	624	955	995	876	741	384	127	5087
Mass.	Boston	40.0	6	60	316	603	983	1088	972	846	513	208	5634
	Pittsfield	32.6	−5	219	524	831	1231	1339	1196	1063	660	326	7578
	Worcester	34.7	−3	147	450	774	1172	1271	1123	998	612	304	6969

State	City	Avg. Winter Temp	Design Temp	Sep	Oct	Nov	Dec	Jan	Feb	Mar	Apr	May	Yearly Total
Mich.	Alpena	29.7	−5	273	580	912	1268	1404	1299	1218	777	446	8506
	Detroit	37.2	4	87	360	738	1088	1181	1058	936	522	220	6232
	Escanaba	29.6	−7	243	539	924	1293	1445	1296	1203	777	456	8481
	Flint	33.1	−1	159	465	843	1212	1330	1198	1066	639	319	7377
	Grand Rapids	34.9	2	135	434	804	1147	1259	1134	1011	579	279	6894
	Lansing	34.8	2	138	431	813	1163	1262	1142	1011	579	273	6909
	Marquette	30.2	−8	240	527	936	1268	1411	1268	1187	771	468	8393
	Muskegon	36.0	4	120	400	762	1088	1209	1100	995	594	310	6696
	Sault Ste. Marie	27.7	−12	279	580	951	1367	1525	1380	1277	810	477	9048
Minn.	Duluth	23.4	−19	330	632	1131	1581	1745	1518	1355	840	490	10000
	Minneapolis	28.3	−14	189	505	1014	1454	1631	1380	1166	621	288	8382
	Rochester	28.8	−17	186	474	1005	1438	1593	1366	1150	630	301	8295
Miss.	Jackson	55.7	21	0	65	315	502	546	414	310	87	0	2239
	Meridian	55.4	20	0	81	339	518	543	417	310	81	0	2289
	Vicksburg	56.9	23	0	53	279	462	512	384	282	69	0	2041
Mo.	Columbia	42.3	2	54	251	651	967	1076	874	716	324	121	5046
	Kansas City	43.9	4	39	220	612	905	1032	818	682	294	109	4711
	St. Joseph	40.3	−1	60	285	708	1039	1172	949	769	348	133	5484
	St. Louis	43.1	4	60	251	627	936	1026	848	704	312	121	4900
	Springfield	44.5	5	45	223	600	877	973	781	660	291	105	4900
Mont.	Billings	34.5	−10	186	487	897	1135	1296	1100	970	570	285	7049
	Glasgow	26.4	−25	270	608	1104	1466	1711	1439	1187	648	335	8996
	Great Falls	32.8	−20	258	543	921	1169	1349	1154	1063	642	384	7750
	Havre	28.1	−22	306	595	1065	1367	1584	1364	1181	657	338	8700
	Helena	31.1	−17	294	601	1002	1265	1438	1170	1042	651	381	8129
	Kalispell	31.4	−7	321	654	1020	1240	1401	1134	1029	639	397	8191
	Miles City	31.2	−19	174	502	972	1296	1504	1252	1057	579	276	7723
	Missoula	31.5	−7	303	651	1035	1287	1420	1120	970	621	391	8125
Neb.	Grand Island	36.0	−6	108	381	834	1172	1314	1089	908	462	211	6530
	Lincoln	38.8	−4	75	301	726	1066	1237	1016	834	402	171	5864
	Norfolk	34.0	−11	111	397	873	1234	1414	1179	983	498	233	6979
	North Platte	35.5	−6	123	440	885	1166	1271	1039	930	519	248	6684
	Omaha	35.6	−5	105	357	828	1175	1355	1126	939	465	208	6612
	Scottsbluff	35.9	−8	138	459	876	1128	1231	1008	921	552	285	6673
Nev.	Elko	34.0	−13	225	561	924	1197	1314	1036	911	621	409	7433
	Ely	33.1	−6	234	592	939	1184	1308	1075	977	672	456	7733
	Las Vegas	53.5	23	0	78	387	617	688	487	335	111	6	2709
	Reno	39.3	2	204	490	801	1026	1073	823	729	510	357	6332
	Winnemucca	36.7	1	210	536	876	1091	1172	916	837	573	363	6761
N.H.	Concord	33.0	−11	177	505	822	1240	1358	1184	1032	636	298	7383
N.J.	Atlantic City	43.2	14	39	251	549	880	936	848	741	420	133	4812
	Newark	42.8	11	30	248	573	921	983	876	729	381	118	4589
	Trenton	42.4	12	57	264	576	924	989	885	753	399	121	4980
N.M.	Albuquerque	45.0	14	12	229	642	868	930	703	595	288	81	4348
	Raton	38.1	−2	126	431	825	1048	1116	904	834	543	301	6228
	Roswell	47.5	16	18	202	573	806	840	641	481	201	31	3793
	Silver City	48.0	14	6	183	525	729	791	605	518	261	87	3705

State	City	Avg. Winter Temp	Design Temp	Sep	Oct	Nov	Dec	Jan	Feb	Mar	Apr	May	Yearly Total
N. Y.	Albany	34.6	− 5	138	440	777	1194	1311	1156	992	564	239	6875
	Binghamton	36.6	− 2	141	406	732	1107	1190	1081	949	543	229	6451
	Buffalo	34.5	3	141	440	777	1156	1256	1145	1039	645	329	7062
	New York	42.8	11	30	233	540	902	986	885	760	408	118	4871
	Rochester	35.4	2	126	415	747	1125	1234	1123	1014	597	279	6748
	Schenectady	35.4	− 5	123	422	756	1159	1283	1131	970	543	211	6650
	Syracuse	35.2	− 2	132	415	744	1153	1271	1140	1004	570	248	6756
N. C.	Asheville	46.7	13	48	245	555	775	784	683	592	273	87	4042
	Charlotte	50.4	18	6	124	438	691	691	582	481	156	22	3191
	Greensboro	47.5	14	33	192	513	778	784	672	552	234	47	3805
	Raleigh	49.4	16	21	164	450	716	725	616	487	180	34	3393
	Wilmington	54.6	23	0	74	291	521	546	462	357	96	0	2347
	Winston-Salem	48.4	14	21	171	483	747	753	652	524	207	37	3595
N. D.	Bismarck	26.6	−24	222	577	1083	1463	1708	1442	1203	645	329	8851
	Devils Lake	22.4	−23	273	642	1191	1634	1872	1579	1345	753	381	9901
	Fargo	24.8	−22	219	574	1107	1569	1789	1520	1262	690	332	9226
	Williston	25.2	−21	261	601	1122	1513	1758	1473	1262	681	357	9243
Ohio	Akron-Canton	38.1	1	96	381	726	1070	1138	1016	871	489	202	6037
	Cincinnati	45.1	8	39	208	558	862	915	790	642	294	96	4410
	Cleveland	37.2	2	105	384	738	1088	1159	1047	918	552	260	6351
	Columbus	39.7	2	84	347	714	1039	1088	949	809	426	171	5660
	Dayton	39.8	0	78	310	696	1045	1097	955	809	429	167	5622
	Mansfield	36.9	1	114	397	768	1110	1169	1042	924	543	245	6403
	Toledo	36.4	1	117	406	792	1138	1200	1056	924	543	242	6494
	Youngstown	36.8	1	120	412	771	1104	1169	1047	921	540	248	6417
Okla.	Oklahoma City	48.3	11	15	164	498	766	868	664	527	189	34	3725
	Tulsa	47.7	12	18	158	522	787	893	683	539	213	47	3860
Ore.	Astoria	45.6	27	210	375	561	679	753	622	636	480	363	5186
	Eugene	45.6	22	129	366	585	719	803	627	589	426	279	4726
	Medford	43.2	21	78	372	678	871	918	697	642	432	242	5008
	Pendleton	42.6	3	111	350	711	884	1017	773	617	396	205	5127
	Portland	45.6	21	114	335	597	735	825	644	586	405	267	4635
	Roseburg	46.3	25	105	329	567	713	766	608	570	405	267	4491
	Salem	45.4	21	111	338	594	729	822	647	611	417	273	4754
Pa.	Allentown	38.9	3	90	353	693	1045	1116	1002	849	471	167	5810
	Erie	36.8	7	102	391	714	1063	1169	1081	973	585	288	6451
	Harrisburg	41.2	9	63	298	648	992	1045	907	766	396	124	5251
	Philadelphia	41.8	11	60	297	620	965	1016	889	747	392	118	5144
	Pittsburgh	38.4	5	105	375	726	1063	1119	1002	874	480	195	5987
	Reading	42.4	6	54	257	597	939	1001	885	735	372	105	4945
	Scranton	37.2	2	132	434	762	1104	1156	1028	893	498	195	6254
	Williamsport	38.5	1	111	375	717	1073	1122	1002	856	468	177	5934
R. I.	Providence	38.8	6	96	372	660	1023	1110	988	868	534	236	5954
S. C.	Charleston	57.9	26	0	34	210	425	443	367	273	42	0	1794
	Columbia	54.0	20	0	84	345	577	570	470	357	81	0	2484
	Florence	54.5	21	0	78	315	552	552	459	347	84	0	2387
	Greenville-Spartanburg	51.6	18	6	121	399	651	660	546	446	132	19	2980
S. D.	Huron	28.8	−16	165	508	1014	1432	1628	1355	1125	600	288	8223

State	City	Avg. Winter Temp	Design Temp	Sep	Oct	Nov	Dec	Jan	Feb	Mar	Apr	May	Yearly Total
	Rapid City	33.4	− 9	165	481	897	1172	1333	1145	1051	615	326	7345
	Sioux Falls	30.6	−14	168	462	972	1361	1544	1285	1082	573	270	7839
Tenn.	Bristol	46.2	11	51	236	573	828	828	700	598	261	68	4143
	Chattanooga	50.3	15	18	143	468	698	722	577	453	150	25	3254
	Knoxville	49.2	13	30	171	489	725	732	613	493	198	43	3494
	Memphis	50.5	17	18	130	447	698	729	585	456	147	22	3232
	Nashville	48.9	12	30	158	495	732	778	644	512	189	40	3578
Tex.	Abilene	53.9	17	0	99	366	586	642	470	347	114	0	2624
	Amarillo	47.0	8	18	205	570	797	877	664	546	252	56	3985
	Austin	59.1	25	0	31	225	388	468	325	223	51	0	1711
	Corpus Christi	64.6	32	0	0	120	220	291	174	109	0	0	914
	Dallas	55.3	19	0	62	321	524	601	440	319	90	6	2363
	El Paso	52.9	21	0	84	414	648	685	445	319	105	0	2700
	Galveston	62.2	32	0	6	147	276	360	263	189	33	0	1274
	Houston	61.0	28	0	6	183	307	384	288	192	36	0	1396
	Laredo	66.0	32	0	0	105	217	267	134	74	0	0	797
	Lubbock	48.8	11	18	174	513	744	800	613	484	201	31	3578
	Port Arthur	60.5	29	0	22	207	329	384	274	192	39	0	1447
	San Antonio	60.1	25	0	31	204	363	428	286	195	39	0	1546
	Waco	57.2	21	0	43	270	456	536	389	270	66	0	2030
	Wichita Falls	53.0	15	0	99	381	632	698	518	378	120	6	2832
Utah	Milford	36.5	− 1	99	443	867	1141	1252	988	822	519	279	6497
	Salt Lake City	38.4	5	81	419	849	1082	1172	910	763	459	233	6052
Vt.	Burlington	29.4	−12	207	539	891	1349	1513	1333	1187	714	353	8269
Va.	Lynchburg	46.0	15	51	223	540	822	849	731	605	267	78	4166
	Norfolk	49.2	20	0	136	408	698	738	655	533	216	37	3421
	Richmond	47.3	14	36	214	495	784	815	703	546	219	53	3865
	Roanoke	46.1	15	51	229	549	825	834	722	614	261	65	4150
Wash.	Olympia	44.2	21	198	422	636	753	834	675	645	450	307	5236
	Seattle	46.9	28	129	329	543	657	738	599	577	396	242	4424
	Spokane	36.5	− 2	168	493	879	1082	1231	980	834	531	288	6655
	Walla Walla	43.8	12	87	310	681	843	986	745	589	342	177	4805
	Yakima	39.1	6	144	450	828	1039	1163	868	713	435	220	5941
W. Va.	Charleston	44.8	9	63	254	591	865	880	770	648	300	96	4476
	Elkins	40.1	1	135	400	729	992	1008	896	791	444	198	5675
	Huntington	45.0	10	63	257	585	856	880	764	636	294	99	4446
	Parkersburg	43.5	8	60	264	606	905	942	826	691	339	115	4754
Wisc.	Green Bay	30.3	−12	174	484	924	1333	1494	1313	1141	654	335	8029
	La Crosse	31.5	−12	153	437	924	1339	1504	1277	1070	540	245	7589
	Madison	30.9	− 9	174	474	930	1330	1473	1274	1113	618	310	7863
	Milwaukee	32.6	− 6	174	471	876	1252	1376	1193	1054	642	372	7635
Wyo.	Casper	33.4	−11	192	524	942	1169	1290	1084	1020	657	381	7410
	Cheyenne	34.2	− 6	219	543	909	1085	1212	1042	1026	702	428	7381
	Lander	31.4	−16	204	555	1020	1299	1417	1145	1017	654	381	7870
	Sheridan	32.5	−12	219	539	948	1200	1355	1154	1051	642	366	7680

Source: *The Solar Home Book,* Anderson and Riordan

Canada Province and City	Sep	Oct	Nov	Dec	Jan	Feb	Mar	Apr	May	Yearly Total
Alberta										
Calgary	410	710	1110	1430	1530	1350	1200	770	460	9520
Edmonton	440	750	1220	1660	1780	1520	1290	760	410	10320
Grande Prairie	450	800	1300	1750	1820	1600	1380	830	460	11010
Lethbridge	350	620	1030	1330	1450	1290	1120	690	400	8650
McMurray	520	880	1500	2070	2210	1820	1540	920	500	12570
Medicine Hat	300	600	1070	1440	1590	1380	1130	620	320	8650
British Columbia										
Atlin	560	870	1240	1590	1790	1540	1370	960	670	11710
Bull Harbour	340	490	630	770	820	710	690	580	470	6370
Crescent Valley	330	680	990	1220	1360	1080	940	610	400	8040
Estevan Point	310	460	580	710	760	670	700	580	470	6090
Fort Nelson	460	920	1680	2190	2200	1870	1460	890	460	12690
Kamloops	200	540	890	1170	1320	1050	780	450	210	6730
Penticton	200	520	820	1050	1190	960	780	490	260	6410
Prince George	460	750	1110	1440	1570	1320	1110	740	480	9720
Prince Rupert	340	510	680	860	910	810	790	650	500	6910
Vancouver	220	440	650	810	890	740	680	480	320	5520
Victoria	260	470	660	790	870	720	690	520	370	5830
Manitoba										
Brandon	350	730	1290	1810	2010	1730	1440	820	420	10930
Churchill	710	1110	1660	2240	2590	2320	2150	1580	1130	16910
Dauphin	320	670	1250	1740	1940	1670	1430	830	420	10560
The Pas	440	840	1480	1980	2200	1850	1620	1010	550	12460
Winnipeg	311	686	1255	1778	1993	1714	1441	810	411	10658
New Brunswick										
Bathurst	310	650	1010	1480	1690	1520	1300	880	520	9670
Chatham	270	640	970	1450	1620	1450	1250	850	490	9290

Canada Province and City	Sep	Oct	Nov	Dec	Jan	Feb	Mar	Apr	May	Yearly Total
Fredericton	250	600	940	1410	1570	1410	1180	780	420	8830
Grand Falls	330	660	1000	1540	1750	1570	1340	870	480	9950
Moncton	260	590	910	1340	1520	1380	1190	830	480	8830
Saint John	280	590	880	1300	1440	1310	1160	830	510	8740
Newfoundland										
Cape Race	350	600	800	1080	1240	1170	1150	950	780	9290
Corner Brook	320	640	890	1200	1410	1360	1240	900	640	9180
Gander	320	660	920	1230	1430	1320	1270	970	650	9440
Goose Bay	440	840	1220	1740	2020	1710	1530	1101	770	12140
St. John's	320	610	820	1130	1270	1180	1170	920	700	8940
Nova Scotia										
Halifax	190	469	745	1109	1262	1180	1042	765	484	7585
Sydney	220	510	780	1130	1310	1280	1160	850	570	8220
Yarmouth	230	480	720	1040	1180	1100	1010	750	510	7520
Ontario										
Fort William	370	740	1170	1680	1830	1580	1380	890	540	10640
Hamilton	140	470	800	1150	1260	1190	1020	670	330	7150
Kapuskasing	420	790	1280	1770	2030	1750	1550	1030	600	11750
Kenora	320	710	1270	1800	1980	1670	1420	860	430	10740
Kingston	160	500	820	1250	1420	1290	1110	710	380	7810
Kitchener	170	520	860	1240	1350	1240	1080	680	330	7620
London	150	490	840	1200	1320	1210	1040	650	330	7380
North Bay	320	670	1080	1550	1710	1530	1350	840	470	9880
Ottawa	200	580	970	1460	1640	1450	1220	730	330	8740
Peterborough	180	540	890	1320	1470	1330	1130	690	1330	8040
Sault Ste. Marie	340	650	1010	1410	1590	1500	1310	820	470	9590
Sioux Lookout	390	780	1310	1850	2060	1750	1510	950	520	11530

Canada Province and City	Sep	Oct	Nov	Dec	Jan	Feb	Mar	Apr	May	Yearly Total
Southampton	190	500	830	1200	1350	1270	1140	760	450	8020
Sudbury	310	680	1100	1500	1720	1450	1340	870	510	9870
Timmins	410	780	1270	1740	1990	1680	1530	1010	550	11480
Toronto	154	465	777	1126	1249	1147	1018	646	316	7008
Trenton	160	470	840	1280	1400	1280	1080	670	330	7630
White River	440	820	1270	1770	1990	1740	1550	1010	590	11850
Windsor	120	410	780	1130	1220	1100	950	580	270	6650
Prince Edward Island										
Charlottetown	240	550	850	1210	1460	1370	1220	870	560	8710
Quebec										
Bagotville	370	740	1160	1730	1950	1710	1450	940	570	11040
Fort Chimo	700	1040	1440	2010	2410	2170	1920	1460	1010	15600
Fort George	550	890	1270	1880	2340	2090	1950	1330	920	14480
Knob Lake	670	1080	1500	2010	2410	2040	1810	1300	910	14890
Megantic	330	660	1000	1480	1640	1490	1290	870	500	9670
Mont Joli	310	660	1030	1440	1650	1470	1310	910	550	9750
Montreal	180	530	890	1370	1540	1370	1150	700	300	8130
Nitchequon	590	970	1430	2050	2340	2010	1820	1310	910	14510
Port Harrison	730	1050	1430	2050	2470	2290	2190	1610	1140	16880
Quebec	250	610	990	1470	1640	1460	1250	810	400	9070
Sherbrooke	240	590	920	1400	1560	1410	1190	750	370	8610
Three Rivers	250	610	980	1490	1690	1490	1250	770	370	9060
Saskatchewan										
North Battleford	380	750	1350	1820	1990	1710	1440	800	400	11000
Prince Albert	410	780	1350	1870	2060	1750	1500	850	440	11430
Regina	370	750	1290	1740	1940	1680	1420	790	420	10770
Saskatoon	380	760	1320	1790	1790	1710	1440	800	420	10960

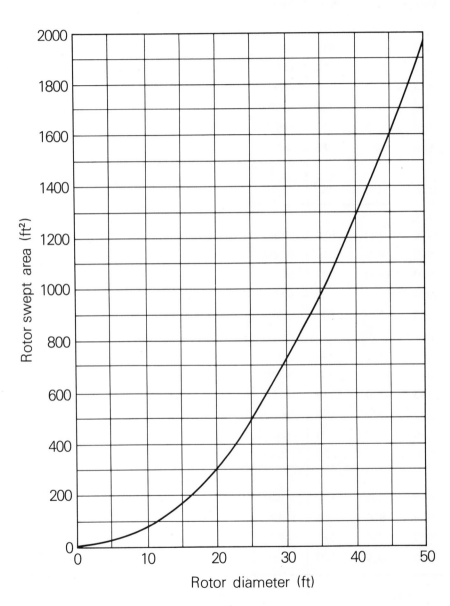

Rotor Swept Area

The rotor swept area, or frontal area, is a parameter frequently used in wind power calculations. This area, denoted by the letter "A," is the total surface area perpendicular to the wind direction that is swept by the rotor blades. It is measured in square feet or square meters.

The first illustration on the facing page indicates the swept area for a propellor-type rotor. This category includes all rotors with a horizontal axis parallel to the windstream; they sweep out an area perpendicular to the wind direction equal to:

$$A = \frac{\pi}{4} \times D^2 = 0.785 \times D^2 ,$$

where $\pi = 3.14159$, and D = diameter is measured from blade tip to blade tip as shown. A convenient graph presented at left will help you convert diameter measurements into rotor swept area, or vice versa.

For vertical-axis, cross-wind machines similar to the second illustration (i.e., those with a uniform radius about the axis of rotation), the swept area is:

$$A = \text{Height} \times \text{Width} .$$

This is the formula to use with Savonius and straight-bladed Darrieus rotors.

In an eggbeater-style Darrieus, the blades assume the shape of a *troposkein* — a complex mathematical curve involving elliptic integrals. Fortunately, this shape can be approximated fairly well with a parabola, and the swept area is about equal to:

$$A = 2.67 \times \text{Radius} \times \text{Half-Height} .$$

The radius and half-height of a typical eggbeater Darrieus are indicated in the third illustration.

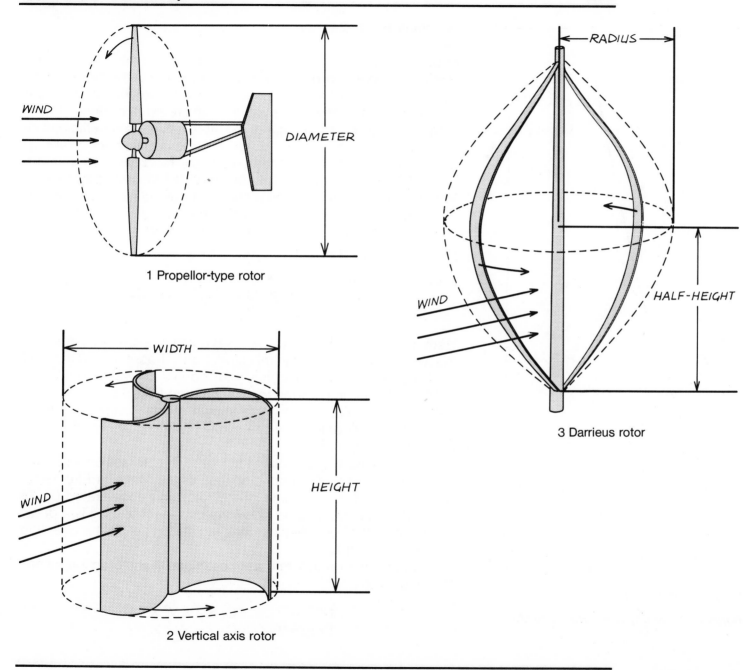

WIND →

DIAMETER

1 Propellor-type rotor

RADIUS

WIND

HALF-HEIGHT

3 Darrieus rotor

WIDTH

WIND

HEIGHT

2 Vertical axis rotor

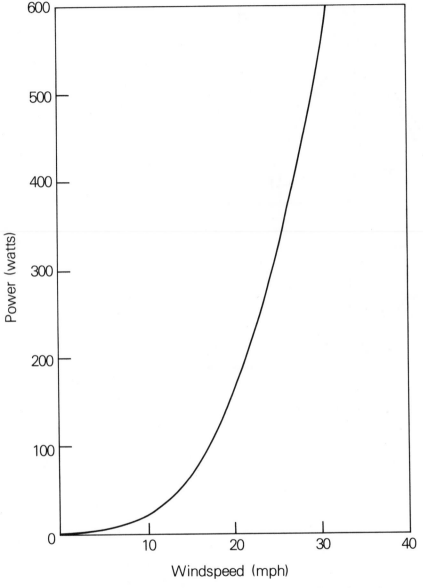

Power curve for a typical Savonius rotor.

Savonius Rotor Design

Suppose you are designing a three-tier Savonius rotor, similar to the one in the photo on page 74. You need to know the power it can deliver at various windspeeds. To calculate this power, use Equation 1 together with the values of the relevant parameters and dimensions of the machine. For example, say the rotor efficiency is given by the graph on page 95, and you expect it to operate at a tip-speed ratio of TSR = 0.8. Then you would read E = 0.15 from that graph for the efficiency of this machine, and use this number in Equation 1. Suppose also that the height of the rotor is 9 feet and its width (or diameter) is 3 feet. Then the rotor swept area is $A = 3 \times 9 = 27$ square feet. Putting this and other information into Equation 1, you get:

$$\text{Power} = \tfrac{1}{2} \times \rho \times V^3 \times A \times E \times K$$
$$= \tfrac{1}{2} \times 0.0023 \times V^3 \times 27 \times 0.15 \times 4.31$$
$$= 0.020 \times V^3 .$$

Here, you have used K = 4.31 so that power is expressed in watts if the windspeed V is given in miles per hour. Thus, if V = 10 mph, the output power of this Savonius rotor equals 20 watts. Performing similar calculations at windspeeds ranging from 5 to 30 mph, you get a power curve like the one shown at left. A similar series of calculations will determine the powershaft torque over the same range of windspeeds.

   You should also calculate the lift and drag forces on an S-rotor. Lift is produced by the rotor because of the Magnus Effect—the wind is slowed on one side of the rotor and accelerated on the other. The lift force pushes sideways on the rotor; drag is a downwind force. Many people estimate drag and forget lift; this oversight could be disastrous.

   Use the following formulas to estimate lift and drag forces on a Savonius rotor:

$$\text{Lift} = 1.08 \times C_L \times \rho \times V^2 \times A ;$$
$$\text{Drag} = 1.08 \times C_D \times \rho \times V^2 \times A ,$$

where C_L is the lift coefficient (see Chapter 5) and C_D is the drag coefficient of the rotor. The parameters ρ, V and A are the usual air density, windspeed and rotor swept area.

Using the accompanying graph of lift and drag coefficients, you can estimate lift and drag forces on this S-rotor. At a TSR = 0.8, the lift coefficient is $C_L = 1.4$ and the drag coefficient is $C_D = 1.2$. Thus, at a windspeed of 20 mph, the lift and drag on this rotor are:

$$\text{Lift} = 1.08 \times 1.4 \times 0.0023 \times 20^2 \times 27$$
$$= 37.6 \text{ pounds ;}$$

$$\text{Drag} = 1.08 \times 1.2 \times 0.0023 \times 20^2 \times 27$$
$$= 32.2 \text{ pounds .}$$

The total force on the Savonius is the vector sum of the lift and drag forces. To calculate the total force, use the following formula:

$$\text{Force} = \sqrt{(\text{Lift})^2 + (\text{Drag})^2} \quad .$$

Then, in our case,

$$\text{Force} = \sqrt{(37.6)^2 + (32.2)^2}$$
$$= \sqrt{2450.6}$$
$$= 49.5 \text{ pounds .}$$

This force is the total load on the support structure. Note that both lift and drag increase with the square of the windspeed—double the windspeed means quadruple the force. In our example, then, the rotor would experience a total force of 198 pounds in a 40 mph wind.

Propellor-type rotors usually employ some kind of governor to prevent the machine from encountering such high forces. But Savonius rotors are difficult to govern; if you apply a brake to slow the rate of rotation, the torque produced by the rotor increases and fights the brake. Moving the vanes so that the S-rotor becomes a cylinder with no exposed vane surface might work. Some people design the tower support system to hold the rotor up in the highest expected wind and just hope for the best.

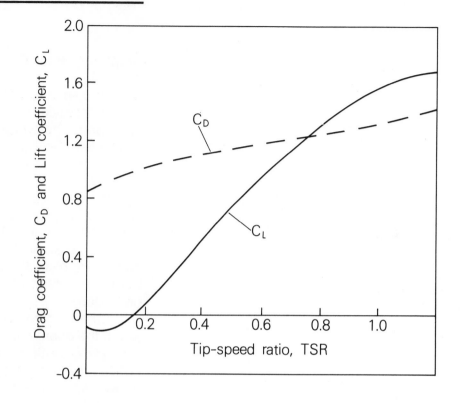

Simplified Rotor Design

There are several methods of sizing a rotor/generator system to supply the annual energy you need. The twelve-step design procedure presented here is a "halfway" method for designing a high-speed propellor-type wind generator. It is much more detailed than the simple procedure given in Chapter 5, but less rigorous than the approaches outlined in some of the references cited in the Bibliography. As long as the aerodynamic sophistication required is not too high, this method is a reliable design procedure.

Start by determining the windspeed at which the energy distribution curve for your site hits a peak. See Chapter 3 for more details on measuring such a distribution. Allow this peak-energy windspeed to be your first cut at a rated windspeed—the speed at which your machine attains its rated power. By this choice, you are approximately centering the machine's prime power-producing range under the wind resource curve. Later refinements in your design calculations may be necessary, but this first pass should give you a wind machine fairly well matched to the site.

The design procedure will be illustrated by an example of a typical small wind generator. Suppose you have an electrical generator with a rated power output equal to 1000 watts. You have monitored the winds at your site and the peak of the energy distribution occurs at 15 mph—a low wind site. From the last table in Chapter 4, you estimate the overall system efficiency of the machine to be about 25 percent. The following design example results in a machine with a cut-in windspeed of about 8 mph and a shut-off, or governing, windspeed of about 40 mph. Thus, it will achieve its rated power over the important windspeeds available at this site.

Step 1. Calculate the rotor diameter. Use Equation 1 to equate the generator rated power to the output power of your system at the peak energy windspeed.

In our example, $E = 0.25$, $V = 15$, and $K = 4.31$ for power expressed in watts and windspeed in mph. Assume $C_A = C_T = 1$. Then,

$$1000 = \frac{1}{2} \times 0.0023 \times 15^3 \times A \times 0.25 \times 4.31 \,,$$

$$\text{or} \quad A = \frac{1000}{\frac{1}{2} \times 0.0023 \times 15^3 \times 0.25 \times 4.31}$$

$$A = 239.1 \ \text{ft}^2 \,.$$

Then the diameter of this machine is $D = 17.5$ feet. Had we assumed a higher system efficiency, say 35 percent instead of 25, we would need a smaller rotor, $D = 14.7$ feet, to develop the same rated power. A higher peak-energy windspeed, say 20 mph instead of 15, would also result in a smaller diameter, D = 11.3 feet.

Step 2. Match the rotor/generator to the site winds either by matching the rotor rpm directly or through selection of an appropriate gearbox. Try to get the tip-speed ratio into the range appropriate for your machine.

If a low-speed generator (DC battery-charger type) is selected, try to use the rpm required by the generator at its rated power as the rotor rpm at the peak energy windspeed. In our example, suppose the generator requires 300 rpm to achieve its rated power of 1000 watts. Then 300 rpm could be chosen to be the rotor rpm at 15 mph and no transmission need be used. However, this rpm corresponds to a TSR of 12.5 (See box on page 71 for calculation of TSR), a bit fast for a 17-foot rotor. A gearbox with a 2:1 gear ratio should be used to lower the rotor rpm to 150 and the TSR to about 6—a much more appropriate value.

In a constant-rpm wind machine, such as an AC device where the rotor drives an induction motor linked to the house wiring, the generator rpm must hold very closely to an rpm dictated by the generator synchronous speed. Most induction motors spin at an rpm of 1800 to 1850. With a 6:1 gearbox, the rotor rpm in our example would be about

300 (TSR = 12.5). To achieve 150 rpm (TSR = 6.2), a 12:1 gear ratio is required.

Step 3. Having selected the design TSR by appropriate gearbox selection, next determine the rotor solidity and the number of blades from the graph and tables on pages 95-96.

At a TSR = 6, a solidity of 0.05 and a three-bladed rotor are the best choices for our example.

Step 4. Draw a rough sketch of the blade, and divide it into several sections. Calculate the speed ratio of the blade at each section by multiplying the design TSR times the fraction of the total distance from hub to the section.

The speed ratio at any radius equals the TSR times that radius divided by the total blade radius, which is 8.75 feet (17.5 ÷ 2) in our example. Thus,

$$SR_A = 6 \times 6 \div 8.75 = 4.11 ,$$

$$SR_B = 6 \times 3 \div 8.75 = 2.06 ,$$

$$SR_C = 6 \times 2 \div 8.75 = 1.37 .$$

For the sake of clarity, only three radius stations plus the tip station are used in this example. Stations every 10 percent of the spar are commonly used in actual design calculations.

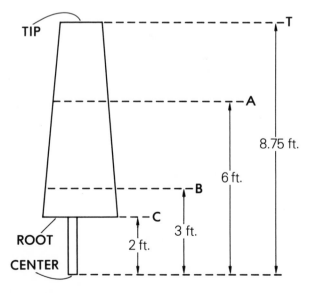

Step 5. From the following graph, read appropriate values of the wind angle φ (Greek "phi") for each station.

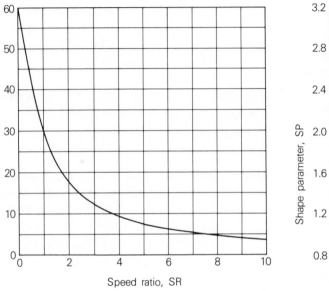

In our example,

$$\phi_T(6.00) = 6.3° ,$$

$$\phi_A(4.11) = 9.0° ,$$

$$\phi_B(2.06) = 17.6° ,$$

$$\phi_C(1.37) = 24.1° .$$

Step 6. From the following graph, read values of the Shape Parameter SP for each station.

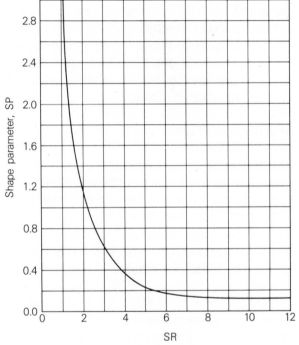

In our example,

$$SP_T = 0.17 ,$$

$$SP_A = 0.37 ,$$

$$SP_B = 1.10 ,$$

$$SP_C = 2.60 .$$

Step 7. Choose an airfoil type and determine the lift coefficient C_L that produces maximum lift-to-drag ratio L/D. Use the graphical method from the box on page 69 of Chapter 4.

Suppose we try the FX60-126 airfoil as used in that box. At the maximum lift-to-drag ratio, L/D = 100, this airfoil has a lift coefficient C_L = 1.08 at an angle of attack equal to 5 degrees. Repeat this step for as many airfoils as you care to try, using drag polar plots from the references. We continue here with the FX60-126.

Step 8. At each radius station, calculate chord lengths using the formula:

$$c = \frac{r \times SP}{C_L \times N},$$

where r is the radius at each station and N is the number of blades.

If r is expressed in feet, then the chord lengths c will be given in feet; multiply by 12 to get your answer in inches. In our example:

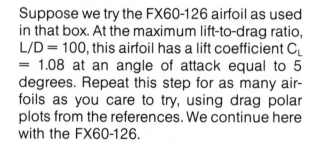

$$c_T = \frac{8.75 \times 0.17}{1.08 \times 3} = 0.46 \text{ ft} (= 5.5 \text{ in}),$$

$$c_A = \frac{6 \times 0.37}{1.08 \times 3} = 0.69 \text{ ft} (= 8.2 \text{ in}),$$

$$c_B = 1.02 \text{ ft} (= 12.2 \text{ in}),$$

$$c_C = 1.60 \text{ ft} (=19.3 \text{ in}).$$

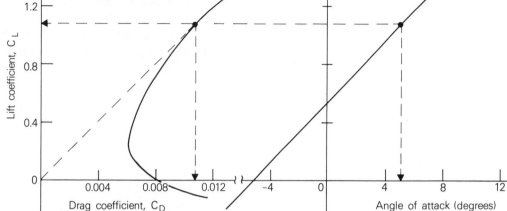

Drag polar plot for the FX60-126 airfoil.

Step 9. Calculate the blade aspect ratio AR, which is just the blade span divided by the average chord length.

From Step 8, the average chord length c is:

$$c = \frac{0.46 + 0.69 + 1.02 + 1.60}{4} = 0.94 \text{ ft}.$$

The blade span, or total radius, is 8.75 feet, so the aspect ratio is,

$$AR = \frac{8.75}{0.94} = 9.3.$$

Step 10. The lift coefficient curve must now be corrected for the blade aspect ratio. This empirical correction adjusts the angle of attack for optimum L/D. Start with the angle of attack a_o where the lift curve crosses the x-axis. Then the corrected angle of attack a_c is given by the formula:

$$a_c = a_o + \frac{C_L}{0.11} \times \left(1 + \frac{3}{AR}\right),$$

where C_L is the lift coefficient at optimum L/D and AR is the blade aspect ratio from Step 9.

In our example, $C_L = 1.08$ and AR = 9.3. From the drag polar plot on page 223, note that $a_o = -5°$, approximately. Thus,

$$a_c = -5 + \frac{1.08}{0.11} \times \left(1 + \frac{3}{9.3}\right)$$

$$= -5 + 9.82 \times 1.3225$$

$$= -5 + 13 = 8°.$$

For more information on this aspect ratio correction, consult *Aircraft Design,* by K. D. Wood.

Step 11. With this corrected angle of attack a_c, calculate the blade angle θ (Greek "theta") from the formula,

$$\theta = \phi - a_c,$$

where the wind angle ϕ is taken from Step 5.

In our example, $a_c = 8.0°$, so ,

$$\theta_T = 6.3 - 8.0 = -1.7°,$$

$$\theta_A = 9.0 - 8.0 = 1.0°,$$

$$\theta_B = 17.6 - 8.0 = 9.6°,$$

$$\theta_C = 24.1 - 8.0 = 16.1°.$$

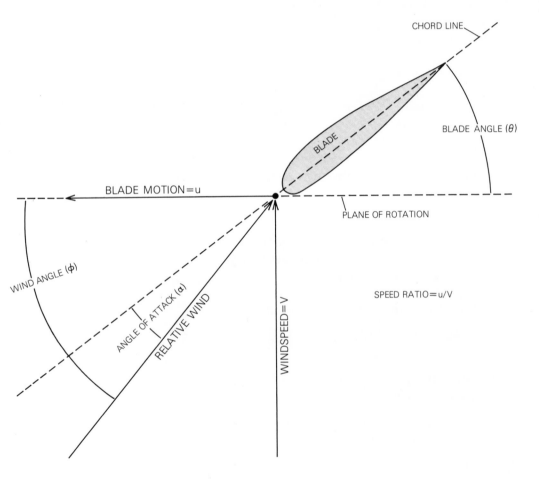

Step 12. Now make an accurate drawing of the blade showing the airfoil cross-sections at their proper blade angles at each station, with the chord lengths calculated in Step 8.

At this point, you can experiment with design variations which include different airfoils and simplifying assumptions like linear twist and taper. The final drawing is the starting point for your structural design efforts.

In general, you will need more detailed information on airfoil sections and performance characteristics than has been presented here. Abbott and Von Doenhoff's *Theory of Wing Sections* contains data on all major NACA airfoils. The January, 1964, and November, 1973, issues of *Soaring Magazine* have data on high-performance Wortmann airfoils. See also the *Handbook of Airfoil Sections for Light Aircraft,* by M. S. Rice. All these references are included in the Bibliography.

Water Consumption Tables

In order to determine the appropriate size of the pump and rotor in a wind-powered water-pumping system, you should first estimate the flow rates likely to be required. These estimates will depend upon the intended uses for the water and the frequency of use. But only rough estimates are practical because different users may have very different habits of water use.

On the average, household water use amounts to 100–150 gallons per person per day—not including outdoor uses like watering lawns and gardens or washing cars. With conservation measures, household use can be cut to perhaps 75 gallons per person per day. Washing a car can take an additional 100 gallons. Depending upon the dryness of climate, lawn-watering can require up to 100 gallons per day for each 1000 square feet of lawn.

Farm water use is more complex. The amount of water used for drinking purposes, for example, will depend on the number and type of animals, their age, the air temperature, and the feed or pasturage being used for them. Average water requirements of cattle, chickens, pigs, sheep and turkeys are listed in the tables in this Appendix. A better treatment can be found in *Wind Power for Farms, Homes and Small Industry*, by Jack Park and Dick Schwind (see Bibliography). Even more thorough tables can be found in the *Yearbook of Agriculture* (1955), issued by the U.S. Department of Agriculture. Consult this book for irrigation water use, which depends on type of crop, soil conditions, climate, and watering practices used.

WATER CONSUMPTION OF LIVESTOCK

Animal	Conditions	Gallons per Day
Holstein Calves	4 weeks old	1.2-1.4
	8 weeks old	1.6
	12 weeks old	2.2-2.4
	16 weeks old	3.0-3.4
	20 weeks old	3.8-4.3
	26 weeks old	4.0-5.8
Dairy Heifers	Pregnant	7.2-8.4
Steers	Maintenance ration	4.2
	Fattening ration	8.4
Jersey Cows	Milk production ½-3½ gal/day	7.2-12.2
Holstein Cows	Milk production 2½-6 gal/day	7.8-21.9
	Milk production 10 gal/day	22.8
	Dry	10.8
Piglets	Body weight = 30 lbs.	0.6-1.2
	Body weight = 60-80 lbs.	0.9
	Body weight = 75-125 lbs.	2.0
	Body weight = 200-380 lbs.	1.4-3.6
Sows	Pregnant	3.6-4.6
	Lactating	4.8-6.0
Sheep	On range or dry pasture	0.6-1.6
	On range (salty feeds)	2.0
	On rations of hay or grain	0.1-0.7
	On good pasture	Very little

WATER CONSUMPTION OF FOWL (per 100 birds)

Animal	Conditions	Gallons per Day
Chicks	1-3 weeks old	0.4-2.0
	3-6 weeks old	1.4-3.0
	6-10 weeks old	3.0-4.0
	9-13 weeks old	4.0-5.0
Pullets	—	3.0-4.0
Hens	Nonlaying	5.0
	Laying (moderate temperatures)	5.0-7.5
	Laying (temperature 90° F)	9.0
Turkeys	1-3 weeks of age	8-18
	4-7 weeks of age	26-59
	9-13 weeks of age	62-100
	15-19 weeks of age	117-118
	21-26 weeks of age	95-105

Estimating Electrical Loads

Your needs for electrical energy and power depend entirely upon the appliances you have and how you use them. To determine these needs, you must examine all your appliances—including electric heaters, power tools and lights—and monitor or estimate their use patterns. First determine the power drawn by each appliance—usually expressed in watts or horsepower and written on a data label somewhere on the appliance. If the appliance is not so labeled, try using an AC wattmeter or consult the manufacturer to determine the rated power. A table of power ratings for typical appliances is included at the end of this Appendix. This table is intended as a rough guide; more accurate power estimates require specific information about the appliances you are using.

If all the appliances were "on" at the same time, the power drawn would be the sum of the rated power levels of all the appliances. In addition, electric motors in appliances like refrigerators and freezers draw up to five times their rated power for a few seconds while starting. Such a surge doesn't contribute very much to the energy demands, but it might put an excessive load on the batteries, inverters or other delicate system components.

In real applications, all the electric devices will rarely, if ever, be on at the same time. The maximum (or "peak") power that your system will have to deliver will be the sum of the rated power levels of all the major devices that are likely to be on at the same time. Accurate load estimates therefore require that you know or can estimate the times when each device is in use. For devices such as television sets, toasters, hot curlers, lights and kitchen appliances, this estimate should be easy: you or your family determine how long they're on.

For other devices like refrigerators, freezers, and electric water heaters, a thermostat controls the on-time or "duty cycle." Some of the methods of determining this duty cycle include:

- Guessing
- Personally recording the on-time with a stopwatch
- Monitoring the on-time with an electric recording device.

The first method is cheap but accuracy can be low. The second is extremely time-consuming and downright boring. The third is accurate but expensive. Guessing can be fairly accurate if the appliance is affected more by your use than by its environment. For example, a refrigerator will draw most of its power around mealtimes —when people are opening it frequently. An electric water heater will draw power after baths, shower, dishwashing and other activities that use a lot of hot water. If you confine your recording activities to such periods, you can still get a pretty accurate estimate of the total time an appliance is in use.

Once you know the rated power of an appliance, and also have a fair estimate of how many hours per month it's being used, you can calculate its monthly energy use by the formula:

$$\text{Energy} = \text{Power} \times \text{Time},$$

where power is expressed in watts and time in hours per month. Divide by 1000 to get your answer in kWh per month. Do this for every appliance in your electrical system and add all the results to get the total electrical energy used per month in kilowatt-hours. As a check, compare the total with your monthly electricity bill. If you're off by more than 25 percent, you might try to adjust some of your estimates and try again.

The next concern is to estimate the peak power required by your electrical system. In normal households linked to the utility lines, the peak demand for electric power usually occurs in the morning near breakfast time or in the evening around dinner. The refrigerator, many lights, several kitchen appliances, and even the TV set are all on at the same time—drawing perhaps a total of 1500 watts. Suddenly, the microwave oven is started and a power spike of 3000 watts occurs for about a half hour. Such a load history is typical of the demands placed on utility lines across the country. Little thought is given to *load management* because old Edison has made provisions to supply this peaking power with extra generators, called "peaking plants" that spring into action as they are needed.

In a wind-electric system, such peak loads can draw too much current and damage a battery bank or delicate system controls. Perhaps a gasoline generator can be used instead to supply this peak power. But with only a little forethought and care, your electric loads can be scheduled to prevent giant power spikes. The microwave oven could be left off, or lights and TV set restricted to minimum use near mealtimes. On the whole, such "peaking shaving" practices greatly enhance the system performance. Some energy will be saved because lower currents in the wires mean lower line losses. Batteries will last longer, and a peaking plant might not be needed at all.

POWER AND ENERGY REQUIREMENTS OF ELECTRIC APPLIANCES			
Item (with Horsepower)	**Watts**	**Hours/Month**	**kWh/Month**
Household Appliances			
Air conditioner, central	—		620
Air conditioner, window (2)	1,566	74	116
Blanket	190	80	15
Blender	350	3	1
Bottle sterilizer	500	30	15
Bottle warmer	500	6	3
Broiler	1,400	6	8.5
Clock	1 - 10	—	1 - 4
Clothes drier	4,600	20	92
Clothes drier, electric heat (6 ½)	4,856	18	86
Clothes drier, gas heat	325	18	6
Clothes washer	—	—	8.5
Clothes washer, automatic	250	12	3
Clothes washer, conventional	200	12	2
Clothes washer, automatic	500	18	9
Clothes washer, wringer	275	15	4
Clippers	40 - 60	—	0.5
Coffee maker	800	15	12
Coffee percolator	300 - 600	—	3 - 10
Coffee pot	900	10	9
Cooling, attic fan (1/6 - ¾)	124 - 560	—	60 - 90
Cooling, refrigeration	—	—	200 - 500
Corn popper	460 - 650	—	1
Curling iron	10 - 20	—	0.5
Dehumidifier	300 - 500	—	50
Dishwasher	1,200	30	36
Disposal (½)	375	2	1
Disposal	500	6	3

POWER AND ENERGY REQUIREMENTS OF ELECTRIC APPLIANCES			
Item (with Horsepower)	Watts	Hours/Month	kWh/Month
Drill	250	2	5
Electric baseboard heater	10,000	160	1,600
Electrocuter, insect	5 - 250	—	1
Electronic oven	3,000 - 7,000	—	100
Fan, attic (½)	375	65	24
Fan, kitchen	250	30	8
Fan, 8 - 16 inches	35 - 210	—	4 - 10
Food blender	200 - 300	—	0.5
Food warming tray	350	20	7
Footwarmer	50 - 100	—	1
Floor polisher	200 - 400	—	1
Freezer, food, 5 - 30 cu. ft.	300 - 800	—	30 - 125
Freezer, ice cream	50 - 300	—	0.5
Freezer, 15 cu. ft.	440	330	145
Freezer, frost free	440	180	57
Fryer, cooker	1,000 - 1,500	—	5
Fryer, deep fat	1,500	4	6
Frying pan	1,200	12	14.5
Furnace, electric control	10 - 30	—	10
Furnace, oil burner	100 - 300	—	25 - 40
Furnace, blower	500 - 700	—	25 - 100
Furnace, stoker	250 - 600	—	3 - 60
Furnace, fan	—	—	32
Garbage disposal equipment (¼ - ⅓)	190 - 250	—	0.5
Griddle	450 - 1,000	—	5
Grill	650 - 1,300	—	5
Hair drier	200 - 1,200	—	0.5 - 6
Hair drier	400	5	2
Heat lamp	125 - 250	—	2

POWER AND ENERGY REQUIREMENTS OF ELECTRIC APPLIANCES			
Item (with Horsepower)	**Watts**	**Hours/Month**	**kWh/Month**
Heater, aux.	1,320	30	40
Heater, portable	660 - 2,000	—	15 - 30
Heating pad	25 - 150	—	1
Heating pad	65	10	1
Heat lamp	250	10	3
Hi-Fi stereo	—	—	9
Hot plate	500 - 1,650	—	7 - 30
House heating	8,000 - 15,000	—	1,000 - 2,500
Humidifier	500	—	5 - 15
Iron	1,100	12	13
Ironer	1 500	12	18
Knife sharpener	125	—	0.2
Lawnmower	1,000	8	8
Lights, 6 room house in winter	—	—	60
Light bulb	75	120	9
Light bulb	40	120	4.8
Mixer	125	6	1
Mixer, food	50 - 200	—	1
Movie projector	300 - 1,000	—	—
Oil burner	500	100	50
Oil burner (1/3)	250	64	16
Polisher	350	6	2
Post light, dusk to dawn	—	—	35
Projector	500	4	2
Pump, water (3/5)	450	44	20
Radio, console	100 - 300	—	5 - 15
Radio, table	40 - 100	—	5 - 10
Range	850 - 1,600	—	100 - 150
Record player	75 - 100	—	1 - 5

POWER AND ENERGY REQUIREMENTS OF ELECTRIC APPLIANCES			
Item (with Horsepower)	**Watts**	**Hours/Month**	**kWh/Month**
Recorder, tape	100	10	1
Refrigerator	200 - 300	—	25 - 30
Refrigerator, conventional	—	—	83
Refrigerator-freezer	200	150	30
Refrigerator-freezer, 14 cu. ft.	325	290	95
Refrigerator-freezer, frost-free	360	500	180
Roaster	1,320	30	40
Rotisserie	1,400	30	42
Sauce pan	300 - 1,400	—	2 - 10
Sewing machine	30 - 100	—	0.5 - 2
Sewing machine	100	10	1
Shaver	12	—	0.1
Skillet	1,000 - 1,350	—	5 - 20
Skil-Saw	1,000	6	6
Sunlamp	400	10	4
Sunlamp	280	5.4	1.5
TV, black and white (AC)	200	120	24
TV, color (AC)	350	120	42
Toaster	1,250	4	5
Typewriter	30	15	5
Vacuum cleaner	600	10	6
Vaporizer	200 - 500	—	2 - 5
Waffle iron	550 - 1,300	—	1 - 2
Washer, automatic	300 - 700	—	3 - 8
Washer, conventional (AC)	100 - 400	—	2 - 4
Water heater (6)	4,474	90	400
Water heater	1,200 - 7,000	—	200 - 300
Water pump, shallow (½)	35	—	5 - 20
Water pump, deep (⅓ - 1)	250 - 746	—	10 - 60

POWER AND ENERGY REQUIREMENTS OF FARM EQUIPMENT			
Item	**Horsepower**	**Watts**	**Kilowatt-hours**
Barn Equipment			
Barn cleaner	2 - 5	1,500 - 3,750	120 per yr.
Corn, ear crushing	1 - 5	750 - 3,750	5 per ton
Corn, ear shelling	¼ - 2	185 - 1,500	1 per ton
Electric fence	—	7 - 10	7 per month
Ensilage blowing	3 - 5	2,250 - 3,750	½ per ton
Feed grinding	1 - 7 ½	750 - 5,600	½ - 1 ½ per 100 lbs.
Feed mixing	½ - 1	375 - 750	1 per ton
Grain cleaning	¼ - ½	185 - 375	1 per ton
Grain drying	1 - 7 ½	750 - 5,600	5 - 7 per ton
Grain elevating	¼ - 5	185 - 3,750	4 per 1000 bu.
Hay curing	3 - 7 ½	2,250 - 5,600	60 per ton
Hay hoisting	½ - 1	375 - 750	⅓ per ton
Milking, portable	¼ - ½	185 - 375	1 ½ per cow/mo.
Milking, pipeline	½ - 3	375 - 2,250	2 ½ per cow/mo.
Silo unloader	2 - 5	1,500 - 3,750	4 - 8 per ton
Silage conveyor	1 - 3	750 - 2,250	1 - 4 per ton
Stock tank heater	—	200 - 1,500	varies widely
Yard lights	—	100 - 500	10 per mo.
Ventilation	1/6 - ⅓	125 - 250	2 - 6 per day per 20 cows
Milkhouse Equipment			
Milk cooling	½ - 5	375 - 3,750	1 per 100 lbs. milk
Space heater	—	1,000 - 3,000	800 per yr.
Water heater	—	1,000 - 5,000	1 per gal.
Poultry Equipment			
Automatic feeder	¼ - ½	185 - 375	10 - 30 per mo.
Brooder	—	200 - 1,000	½ - 1 ½ per chick per season

POWER AND ENERGY REQUIREMENTS OF FARM EQUIPMENT			
Item	Horsepower	Watts	Kilowatt-hours
Burglar alarm	—	10 - 60	2 per mo.
Debeaker	—	200 - 500	1 per 3 hrs.
Egg cleaning or washing	—	—	1 per 2000 eggs
Egg cooling	1/6 - 1	125 - 750	1 ¼ per case
Night lighting	—	40 - 60	10 per mo. per 100 birds
Ventilating fan	—	50 - 300	1 - 1 ½ per day
Water warming	—	50 - 700	varies widely
Hog Equipment			
Brooding	—	100 - 300	35 per brooding period per litter
Ventilating fan	—	50 - 300	¼ - 1 ½ per day
Water warming	—	50 - 1,000	30 per brooding period per litter
Farm Shop			
Air compressor	¼ - ½	185 - 375	1 per 3 hr.
Battery charging	—	600 - 750	2 per battery charge
Concrete mixing	¼ - 2	185 - 1,500	1 per cu. yd.
Drill press	1/6 - 1	125 - 750	½ per hr.
Fan, 10 - inch	—	35 - 55	1 per 20 hr
Grinding, emery wheel	¼ - ⅓	185 - 250	1 per 3 hr.
Heater, portable	—	1,000 - 3,000	10 per mo.
Heater, engine	—	100 - 300	1 per 5 hr.
Lighting	—	50 - 250	4 per mo.
Lathe, metal	¼ - 1	185 - 750	1 per 3 hr.
Lathe, wood	¼ - 1	185 - 750	1 per 3 hr.
Sawing, circular, 8 - 10 inch	⅓ - ½	250 - 375	½ per hr.
Sawing, jig	¼ - ⅓	185 - 250	1 per 3 hr.
Soldering, iron	—	60 - 500	1 per 5 hr.

POWER AND ENERGY REQUIREMENTS OF FARM EQUIPMENT			
Item	Horsepower	Watts	Kilowatt-hours
Miscellaneous			
Farm chore motors	½ - 5	375 - 3,750	1 per hp per hr.
Insect trap	—	25 - 40	3 per night
Irrigation pumps	1 and up	750 and up	1 per hp per hr.
Wood sawing	1 - 5	750 - 3,750	2 per cord

BIBLIOGRAPHY

Books

Abbott and Von Doenhoff, *Theory of Wing Sections.* New York: Dover Publications, 1959. Contains data on all the major NACA airfoils.

Bruhn, E. F., *Analysis and Design of Flight Vehicle Structures.* Available from Tri-State Offset Company, 817 Main Street, Cincinnati, Ohio 45202. Several completely untrained people have used this book to design rather sophisticated home-built aircraft; a great source of data.

Eldridge, Frank R., *Wind Machines* (Second Edition). New York: Van Nostrand Reinhold Company, 1980. The updated book-length version of Eldridge's MITRE Corporation study of the wind power field; an excellent resource for the professional engineer, with emphasis on large-scale systems.

Fraenkel, Peter L., *Food from Windmills.* Available from Intermediate Technology Publications, Ltd., 9 King Street, London WC2 8HN, United Kingdom.

Golding, E. W., *The Generation of Electricity by Wind Power.* London: The Philosophical Library, 1955; reprinted by the Halstead Press, John Wiley and Sons, Inc., New York, 1976. The classic text on wind-generated electricity.

Hackleman, Michael, *Wind and Windspinners.* Culver City, California: Peace Press, 1974. Also available from Earthmind, Boyer Road, Mariposa, California 95338. A popular how-to book on Savonius rotor design and construction, but its advocacy of their use for electrical generation is somewhat impractical.

Inglis, David R., *Wind Power and Other Energy Options.* Ann Arbor, Michigan: University of Michigan Press, 1978.

Justus, C. G., *Winds and Wind System Performance.* Philadelphia: Franklin Institute Press, 1978.

National Weather Service and Federal Aviation Administration, *Aviation Weather.* Available from Aviation Maintenance Publishers, P. O. Box 890, Basin, Wyoming 82410.

Putnam, Palmer C., *Power from the Wind,* New York: Van Nostrand Reinhold Company, 1974. A reprinted edition of the classic 1948 study on the design and construction of the Smith-Putnam wind generator.

Rice, M. S., *Handbook of Airfoil Sections for Light Aircraft.* Available from Aviation Publications, P. O. Box 123, Milwaukee,

Wisconsin 53201.

Stern, Peter, *Small-Scale Irrigation.* London: Intermediate Technology Publications, Ltd., 1979.

Torrey, Volta, *Wind Catchers: American Windmills of Yesterday and Tomorrow.* Brattleboro, Vermont: Stephen Greene Press, 1976.

Articles and Reports

Barchet, W. R. and D. Elliott, "Northwest Wind Energy Resource Assessment Atlas," 1979. Battelle Pacific Northwest Laboratory Report. Available from National Technical Information Service, U.S. Department of Commerce, Springfield, Virginia 22161 (hereafter referred to as NTIS).

Berry, Edwin X., "Wind Resource Assessment in California." Available from Wind Energy Office, California Energy Commission, 1111 Howe Avenue, Sacramento, California 95825.

Eldridge, Frank R., editor, "Proceedings of the Second Workshop on Wind Energy Conversion Systems," June, 1975. Published by the MITRE Corp., McLean, Virginia; available from NTIS (Document No. NSF-RANN-75-050).

Eldridge, Frank R., "Wind Machines," October, 1975. Available from the Superintendent of Documents, U.S. Government Printing Office, Washington, DC 20402 (Stock No. 038-000-00272-4).

Fales, E. N., "Windmills," in *Standard Handbook for Mechanical Engineers,* T. Baumeister and L. S. Marks, editors. New York: McGraw-Hill Book Company, 1967 (7th edition).

Golding, E. W., "Electrical Energy from the Wind," in *Proceedings of the Institution of Electrical Engineers,* Vol 102A, pp. 667-695 (1955).

Hütter, Ulrich, "Optimum Wind Energy Conversion Systems," in *Annual Reviews of Fluid Mechanics,* Vol. 9, pp. 399-419 (1977).

"Legal-Institutional Implications of Wind Energy Conversion Systems," September, 1977. Available from NTIS (Report No. NSF/RA-770203).

Merriam, Marshall, "Wind Energy for Human Needs," in *Technology Review,* Vol. 79, No. 3, pp. 28-39 (1977).

Park, Jack and Dick Schwind, "Wind Power for Farms, Homes and Small Industry," September, 1978. Available from NTIS (Report No. RFP-2841/1270/78/4).

Rogers, S. E., et al., "An Evaluation of the Potential Environmental Effects of Wind Energy System Development," August, 1976. Battelle Columbus Laboratories Report No. ERDA/NSF/07378-75/1. Available from NTIS.

Reed, Jack, "Wind Power Climatology in the United States," June, 1975. Sandia Laboratories Report No. SAND-74-0348. Available from NTIS.

Reed, Jack, "Wind Power Climatology of the United States, Supplement," 1979. Available from NTIS.

Savino, J. M., editor, "Wind Energy Conversion Systems—Workshop Proceedings," June, 1973. National Science Foundation Document No. NSF/RANN-73-006. Available from NTIS.

Weatherholt, Lyle, editor, "Vertical-Axis Wind Turbine Technology Workshop Proceedings," May, 1976. Sandia Laboratories Report No. SAND 76-5586. Available from NTIS.

Wegley, H. L., et al., "A Siting Handbook for Small Wind Energy Conversion Systems," Battelle Pacific Northwest Laboratories Report No. PNL-2521. Available from NTIS.

Wilson, Robert E. and Peter B. S. Lissaman, "Applied Aerodynamics of Wind Power Machines," July, 1974. Oregon State University Report No. NSF-AER-74-04014. Available from NTIS.

Wilson, Robert E., et al., "Aerodynamic Performance of Wind Turbines," June, 1976. Oregon State University Report No. ERDA/NSF/04014-76/1. Available from NTIS.

"Wind Energy," in *Proceedings of the United Nations Conference on New Sources of Energy,* Rome, 1961, Vol. 7. Published by the United Nations.

"Wind Energy Conversion Systems: Workshop Proceedings," 1977. Available from NTIS (Document No. CONF-770921).

Address for NTIS: National Technical Information Service, U. S. Department of Commerce, Springfield, Virginia 22161

Periodicals

Alternative Sources of Energy, Milaca, Minnesota 56353.

Canadian Renewable Energy News, P.O. Box 4869, Station E, Ottawa, Ontario K1S 5B4, Canada.

RAIN: Journal of Appropriate Technology, 2270 N.W. Irving, Portland, Oregon 97210.

Soaring Magazine, published by Soaring Society of America, P.O. Box 66071, Los Angeles, California 90066.

Solar Age, published by Solarvision, Inc., Church Hill, Harrisville, New Hampshire 03450.

Wind Energy Report, P.O. Box 14, Rockville Center, New York 11571.

Wind Engineering, published by Multi-Science Publishing Company, Ltd, The Old Mill, Dorset Place, London E15 10J, United Kingdom.

Windletter, newsletter of the American Wind Energy Association, 1609 Connecticut Avenue, N.W., Washington, DC 20009.

Wind Power Digest, 115 East Lexington, Elkhart, Indiana 46516.

Wind Technology Journal, published by the American Wind Energy Association, 1609 Connecticut Avenue, N.W., Washington, DC 20009.

Bibliographies

"Energy from the Wind: Annotated Bibliography," by Barbara L. Burke and Robert Meroney, April 1975 plus annual supplements. Available from Colorado State University Engineering Research Center, Publications Department, Fort Collins, Colorado 80523.

Energy Primer, updated and revised edition, by Richard Merrill and Thomas Gage. New York: Delta Books, 1978. The Wind section, pp. 120-149, contains extensive reviews of available books and reports.

"Wind Energy: A Bibliography with Abstracts and Keywords," by R. van Steyn. Available from Library Administration, University of Technology, P.O. Box 513, Eindhoven, Netherlands.

"Wind Energy Bibliography," 1973. Available from Windworks, P.O. Box 329, Mukwonago, Wisconsin.

"Wind Energy Information Directory," October, 1979. Solar Energy Research Institute Report No. 061-000-00350-9. Available from Superintendent of Documents, U.S. Government Printing Office, Washington, DC 20402.

"Wind Energy Information Sources." Available from Document Distribution Service, Solar Energy Research Institute, Golden, Colorado 80401.

"Wind Energy Utilization: A Bibliography," April, 1975. Technology Applications Center Report No. TAC-W-75-700, available from Publications and Documents, University of New Mexico, Albuquerque, New Mexico 87131. A complete bibliography covering the period 1944-1974.

Construction Plans

A. Bodek, "How to Construct a Cheap Wind Machine for Pumping Water," Do-It-Yourself Leaflet No. L-5, Brace Research Institute, MacDonald College, McGill University, Ste. Anne de Bellevue 800, Quebec H9X 3M1, Canada.

R. D. Mann, "How to Build a 'Cretan Sail' Windpump." Available from Intermediate Technology Publications, Ltd., 9 King Street, London WC2E 8HN, United Kingdom.

Helion, Inc., "12/16 Construction Plans," for a 2000-watt wind generator with 12- or 16-foot diameter blades. Order from Helion, Inc., P.O. Box 445, Brownsville, California 95919.

Equipment Directories

"A Guide to Commercially Available Wind Machines," Rocky Flats Wind Systems Program Report No. RFP-2836/3533/78/3; available from NTIS. Contains detailed descriptions of 66 wind machines under 100 kW available in the United States and Canada; includes illustrations and power curves, plus directory of manufacturers and dealers.

Energy Primer, updated and revised edition, by Richard Merrill and Thomas Gage. New York: Delta Books, 1978. The Wind section, pp. 120-148, contains evaluations, dealers and distributors of small-scale wind machines and related components.

Harnessing the Wind for Home Energy, by Dermot McGuigan. Charlotte, Vermont: Garden Way Publishing Company, 1978. Contains descriptions and costs of wind machines and other equipment available in 1978.

The American Wind Energy Association publishes a free brochure that contains, among other things, a list of commercially available wind machines and their U.S. distributors. Write AWEA, 1609 Connecticut Avenue, N.W., Washington DC 20009, or call (202) 667-9137.

GLOSSARY

AC—alternating electric current, which reverses its direction at regular intervals; see also *current.*

airfoil—a curved surface designed to create aerodynamic lift forces when air flows around it.

ampere, or amp—a measure of electric current flow, equal to one coulomb of electric charge per second. Also, the current that flows through a resistance of one ohm when the applied voltage is one volt.

ampere-hour, or **amp-hour**—a measure of electric charge, calculated by multiplying the current (in amperes) by the number of hours it flows; see *ampere.*

anemometer—an instrument used for measuring the windspeed.

asynchronous generator—an electrical generator designed to produce an alternating current that matches an existing power source (e.g., utility lines) so the two sources can be combined to power one load. The generator does not have to turn at a precise rpm to remain at the correct frequency or phase; see also *synchronous generator.*

current—the flow of electric charge through a wire; see also *AC* and *DC*.

cut-in speed—the windspeed at which a wind machine begins to produce power.

cut-out speed—the windspeed at which a wind machine is shut down to prevent high-wind damage; Also called *furling speed*.

Darrieus rotor—a vertical-axis, lift-type rotor, generally with two or three blades, that has high efficiency and low torque.

DC—direct electrical current, which flows in only one direction along a wire.

drag—an aerodynamic force that retards the normal motion of lift-type rotor blades, or actually causes vane motion and produces power in drag-type wind machines.

efficiency—a number, usually expressed as a percentage, equal to the power output of a device divided by the input power to that device; see also *power coefficient*.

energy—a measure of the work that might be, or has been done, usually expressed in kilowatt-hours (kWh) or horsepower-hours (hphr); see also *work*.

fantail—a propellor-type wind rotor mounted sideways on a larger wind machine (horizontal-axis) and used to keep that machine aimed into the wind.

fetch area—the terrain over which the wind usually blows before reaching a wind machine or turbine.

furling speed—the windspeed at which the wind machine must be shut down to prevent high-wind damage; also called *cut-out speed*.

gear ratio—a ratio of the rotational speeds (in rpm) between the rotor powershaft and the pump, generator, or other device powershaft; the term applies both to speed-increasing and speed-decreasing transmissions.

gin pole—a pipe, board or tower used for improved leverage in raising a tower.

head—the total height a pump must lift water.

horsepower (hp)—the standard English system measure of power, equal to 550 foot-pounds per second; see also *power*.

horsepower-hour (hphr)—a measure of energy in the English system; see also *energy* and *horsepower*.

inverter—a device that converts direct current to alternating current; see also *AC, DC,* and *synchronous inverter*.

kilowatt (kW)—a measure of power in the metric system, equal to 1,000 watts; see also *watt* and *power*.

kilowatt-hour (kWh)—a measure of electrical energy in the metric system, equal to 1,000 watt-hours, calculated by multiplying kilowatts times the number of hours; see also *kilowatt, horsepower* and *watt*.

lift—the aerodynamic force that "pulls" a rotor blade along its rotary path.

megawatt—a measure of power in the metric system, equal to one million watts or 1,000 kilowatts; see also *watt* and *kilowatt.*

panemone—a drag-type, vertical-axis wind machine; the term could describe a Darrieus rotor, but is generally used only for drag-type machines.

power—the rate at which work is performed or energy is consumed; mechanical power is equal to force times velocity, while electric power is equal to the voltage times the amperes of current flow.

power coefficient—the ratio of power output to power input, usually for a wind machine rotor; often referred to as the efficiency; see also *efficiency, rotor efficiency.*

rated power—the expected power output of a wind machine, either its maximum power output or an output at some windspeed less than the maximum speed before governing controls start to reduce the power.

rated speed—the windspeed at which a wind machine delivers its rated power; it can be the speed at which a governor takes over or a lower windspeed; see also *rated power.*

resistance—the tendency of almost any electrical device to impede the flow of electrical current.

return time—the time elapsed before the windspeed returns to a higher, specified value, such as the cut-in speed of a windmill or rotor.

rotor—the rotating structure of a wind machine, also called the wind-wheel, that includes blades and powershaft and converts wind power into rotary mechanical power.

rotor efficiency—the efficiency of the rotor alone; it does not include any losses in transmissions, pumps, generators, or line or head losses. Also called rotor power coefficient; see also *efficiency.*

Savonius rotor—a vertical axis, drag-type rotor that generally has low efficiency but high starting torque.

sine wave—the type of alternating current generated by utilities, rotary inverters and AC generators.

solidity—the ratio of the blade surface area in a rotor to the total frontal area (or swept area) of the entire rotor.

square wave—the type of alternating current output from low-cost, solid-state inverters, which is usable for many appliances but may affect stereos and TV sets adversely.

synchronous generator—an alternating-current generator which operates together with another AC power source and must turn at a precise rpm to hold the frequency and phase of its output current equal to those of the AC source; see also *AC, DC, inverter.*

synchronous inverter—an electronic device that converts DC to AC but must have another AC source (e.g., the utility lines) for voltage and frequency reference; the AC output of the device is in phase and at the same frequency as the outside AC source; see also *AC, DC, inverter.*

torque—the moment of force produced by the rotor blades that causes a power-shaft to turn.

turbulence—rapid windspeed fluctuations; gusts are maximum values of wind turbulence.

voltage—the electrical potential, or pressure, that causes current to flow through a wire; it is measured in volts.

watt—a measure of electric power in the metric system, equal to one ampere flowing through a potential difference of one volt.

watt-hour—a measure of electrical energy in the metric system, equal to one watt times one hour; see also *watt, kilowatt-hour.*

wind furnace—a wind system that converts available wind power into useful heat; see also *wind machine.*

wind generator—a wind machine that produces electrical power; see also *wind machine.*

wind machine—a device that extracts useful power from the wind, but mainly one that produces a mechanical or electrical output; also called *wind system* or *wind turbine.*

windmill—an archaic term for wind system that is still used to refer to high-solidity rotors in water-pumpers and older mechanical-output machines.

wind power—the power actually contained in the wind, or the fraction of that power extracted by a wind machine; see also *power.*

wind rose—a two-dimensional graph showing the monthly or yearly average windspeed from each of usually 16 directions and the percentage of time the wind blows from each direction.

windspeed distribution—a two-dimensional graph that shows the total time or the percentage of time that the wind blows at each windspeed, for a given site.

windspeed profile—a two-dimensional graph indicating how the windspeed changes with height above the ground or water.

windwheel—see *rotor.*

work—a form of mechanical energy equal to a force multiplied by the distance moved along the force direction.

yaw axis—the vertical axis about which an entire propellor-type wind machine rotates to align itself with the wind and its rotor perpendicular to the wind; also called *lolly axis.*

INDEX

Page numbers printed in *italics* indicate tables, photographs or illustrations.

Multiblade farm windmills, *14,* 80-83, *80, 81, 82,* 87, *88. See also* American Farm windmill.

N

National Weather Service (NWS), 52-53
New Alchemy Institute, *16*

O

Output power, *70*

P

Palmdale, California, 55-56, *55*
Panemones, 13, 43, 114
Performance curves for S-rotors, *95*
Piston
    compressors, 87
    water-pumps, 87
Pitching moment, 69
Popular Science, 16
Post windmills, 13-14
Power
    coefficients, 71, *95*
    rated, 68
    requirements for electrical appliances, 132-37, *133*
    requirements for water pumping, 123-26, *125*
Powershaft torque, 107
Propellor-type rotors, 74, 83-85, *83,* 86, 87, 95-99
Public Utility Regulatory Policies Act (PURPA), 153, 155

R

Rated power, 68
Rayleigh distribution, 52-55, *53, 54,* 59
Recorders, data, 60-63, *62, 63*
Refrigeration, 142
Residential Conservation Service (RCS), 155
Ropes, 147
Rotor
    coning, 112, *113*
    drag, 106
    efficiency, 98
    power coefficient, 71, *95*
    solidity, 91
Rotors
    design of, 32-34, 88
    high speed, 83-85, 97
    performance curves for, *93*
    sizing, 88
    well-matched to wind characteristics, *98*
    See also Darrieus rotors; Multiblade rotors; Propellor-type rotors; Sailwing water pumpers; Savonius rotors
Rural Electrification Administration (REA), 14, 17, 21, 31

S

Safety hazards, 153-54, *154*
Sailwing water pumper, *16,* 25-28, *26-28,* 87, 88
St. Ann's Head, England, *54*
Sandia Laboratories, *76*

Biographical Note

A native Californian, Jack Park is one of America's leading authorities on small wind-power systems. He has designed and built dozens of small wind machines, including a series of wind generators marketed by Kedco, Inc. He is currently (1981) President of the American Wind Energy Association, having served previously as Vice President and Director. Jack is also the founder and President of Helion, Inc., a consulting, engineering and design firm specializing in solar and wind energy devices. He has traveled throughout the United States, and to parts of Canada, Europe and Africa—speaking to diverse audiences about the small-scale uses of wind power.

Jack has served on the Faculty of Goddard College and the Southern California Institute of Architecture, teaching courses on renewable energy technology and applications. Formerly technical editor of the *Wind Power Digest,* he has contributed many articles to that journal and to *Alternative Sources of Energy* Magazine. He has two previous books to his credit: *Simplified Wind Power Systems for Experimenters,* published by Helion, and *Wind Power for Farms, Homes and Small Industry,* a Department of Energy publication he wrote with Richard Schwind.

When not away lecturing or consulting, Jack lives on his northern California ranch with his wife Helen—and their many windmills and farm animals.

Colophon

This book was designed by Edward Wong-Ligda and Linda Goodman. The text typeface is Helvetica Light and Bold, and was set on a Quadritek photo-typesetter by Abracadabra Typesetting of Palo Alto, California. The paper is Perkins & Squier Finch Offset Vellum. The cover color separation is by Color Tech of Redwood City, California. The printing and binding is by Malloy Lithographing of Ann Arbor, Michigan.